Sustaining Natural Resources in a Changing Environment

Edited by
Linda Hantrais, Ruth Kattumuri and Ashley Thomas Lenihan

Routledge
Taylor & Francis Group
LONDON AND NEW YORK

First published 2019
by Routledge
2 Park Square, Milton Park, Abingdon, Oxon, OX14 4RN, UK

and by Routledge
52 Vanderbilt Avenue, New York, NY 10017, USA

First issued in paperback 2020

Routledge is an imprint of the Taylor & Francis Group, an informa business

British Library Cataloguing in Publication Data
A catalogue record for this book is available from the British Library

ISBN 13: 978-0-367-66377-3 (pbk)
ISBN 13: 978-1-138-36170-6 (hbk)

Typeset in Myriad Pro
by RefineCatch Limited, Bungay, Suffolk

Publisher's Note
The publisher accepts responsibility for any inconsistencies that may have arisen during the conversion of this book from journal articles to book chapters, namely the possible inclusion of journal terminology.

Contents

Sustaining Natural Resources in a Changing Environment

Climate change and environmental degradation have intensified the pressures on crucial resources such as food and water security and air quality. In this collection, academic researchers and practitioners who have lived and worked in countries as geographically and culturally diverse as Brazil, China, India, Ghana, Palestine, Uganda and Venezuela draw on their wide-ranging international and inter-sectoral experience to offer valuable comparative insights into the relationship between research and evidence-based policy for sustaining natural resources. Their contributions provide a novel mix of disciplinary perspectives ranging across geography, ecology, social policy, the political economy, philosophy, international development, engineering technology, architecture and urban planning. They examine the institutions involved in generating and mediating evidence about the sustainability of natural resources in a changing environment, and the different methodologies employed in collecting and assessing evidence, informing policy and contributing to governance. The authors demonstrate not only that social science evidence on governance and policy implementation to sustain natural resources must complement natural science inputs, but also that local communities must be an integral part of any programme development.

This book was originally published as a special issue of *Contemporary Social Science.*

Linda Hantrais, FAcSS, is Emeritus Professor of European Social Policy at Loughborough University, UK, and Visiting Professor at the LSE Centre for International Studies, UK. Her research interests span international comparative research theory, methods, management and practice, with particular reference to public policy and institutional structures, and the relationship between socio-demographic trends and social policy.

Ruth Kattumuri, FAcSS, is Co-Director of the LSE India Observatory and an Associate at the Grantham Research Institute on Climate Change and the Environment, UK. She is involved in transdisciplinary evidence-based research, policy influence and advice concerning sustainable growth, inclusion and population development, primarily in the context of the Global South.

Ashley Thomas Lenihan is a Visiting Fellow at the LSE Centre for International Studies, UK, and a Term Member of the Council on Foreign Relations, USA. Her research focusses on the connection between state power and foreign investment, and the relationship between social science research and the policymaking process.

Citation Information

The chapters in this book were originally published in *Contemporary Social Science*, volume 13, issue 1 (March 2018). When citing this material, please use the original page numbering for each article, as follows:

Chapter 1
Sustaining natural resources in a changing environment: evidence, policy and impact
Ruth Kattumuri
Contemporary Social Science, volume 13, issue 1 (March 2018), pp. 1–16

Chapter 2
International water targets and national realities in Sub-Saharan Africa: the case of Uganda
Kevin Sansom, David Hirst and Sam Kayaga
Contemporary Social Science, volume 13, issue 1 (March 2018), pp. 17–29

Chapter 3
Towards sustainable development: planning surface coal mine closures in India
Raman Srikanth and Hippu Salk Kristle Nathan
Contemporary Social Science, volume 13, issue 1 (March 2018), pp. 30–43

Chapter 4
The impact of urbanisation on land resources in Chongqing, China, 1997–2015
Huiming Zong and Bingjie Cai
Contemporary Social Science, volume 13, issue 1 (March 2018), pp. 44–57

Chapter 5
Negotiating transparency: NGOs and contentious politics of the Extractive Industries Transparency Initiative in Ghana
Nelson Oppong
Contemporary Social Science, volume 13, issue 1 (March 2018), pp. 58–71

Chapter 6
Gendered participation in community forest governance in India
Satyapriya Rout
Contemporary Social Science, volume 13, issue 1 (March 2018), pp. 72–84

Chapter 7

Community empowerment in changing environments: creating value through food security
Nicole Kenton and Sumita Singha
Contemporary Social Science, volume 13, issue 1 (March 2018), pp. 85–99

Chapter 8

Environmental impact assessment: evidence-based policymaking in Brazil
Kathryn Hochstetler
Contemporary Social Science, volume 13, issue 1 (March 2018), pp. 100–111

Chapter 9

Low carbon energy and international development: from research impact to policymaking
Ed Brown, Ben Campbell, Jon Cloke, Long Seng To, Britta Turner and Alistair Wray
Contemporary Social Science, volume 13, issue 1 (March 2018), pp. 112–127

Foreword

It is well established that climate change and environmental degradation are intensifying the pressures on crucial resources such as food and water security and air quality. These pressures are especially difficult to counteract in the least developed economies. Yet if policies are to reduce these pressures they must draw on effective evidence. This evidence needs not only to be drawn from natural scientists who document the impact of such factors as population growth and accelerated urbanisation against the backdrop of global warming, but also from social scientists' examinations of how policies are developed and implemented. Contributions to these pressing issues are revealed in the variety of chapters in the present volume.

The studies in this volume make it clear that there are many difficulties to be overcome if policymaking is to achieve its goals. This is illustrated by Kevin Sansom from Loughborough University and his associates in their study of the provision of piped water to homes in Uganda. They show that an emphasis on international targets, and national attempts to achieve those long-term goals, ignore complications which occur at the local level. Shorter-term objectives developed at the grass roots are much more likely to be achieved.

When carried out effectively, programmes have shown that some environmental degradation is reversible. For example, the impact of opencast coal mines on local forestry and water resources, air quality and biodiversity can be mitigated. But as Raman Srikanth and his colleague from the National Institute of Advanced Studies in Bengaluru discuss, there are many shortcomings in the Indian government's regulatory regimes and policies for doing this. They point out that, whilst there is a good history, in the US and Australia, of restoration once surface coal mines have been closed, the intricacies of the problem are not currently embraced in India. Interestingly the authors of this critique have been entrusted with a research project to explore the feasibility of their proposals. But, intriguingly, although their recommendations involve legal, political and administrative foci, the authors are both engineers.

In contrast to the democratic, some would say chaotic, quality of Indian governance, the strong central government of China can directly impose requirements on the management of environmental resources alongside rapid urbanisation. This is illustrated in great detail by Huiming Zong and Bingjie Cai of China's Southwest University in Chongqing. They show how, for an autocratic government to be effective, close involvement of social scientists with regional and local governments is essential. They argue that, when careful analysis is carried out of the impact of urbanisation, all those involved at every level of government can make effective decisions. In some cases, this has even meant that a rather unusual step was taken of not using arable land for construction.

The Chinese approach does seem to involve local government although it is driven by central edicts. This is crucial as is demonstrated in all these chapters. The challenge is to get grass-roots involvement. Without this little effective change is possible. I am reminded of the parallel to communication in organisations being like the flow of water. It moves effortlessly from the top down but requires special pressure to move up.

Local community involvement, though, does not inevitably overcome all the challenges of environmental change and its related livelihood insecurity. In an examination of participatory bottom-up approaches to forestry governance in India, Satyapriya Rout from the University of Hyderabad shows that the existing social norms can distort the objectives of any programme. In the cases he examines he claims that, although some aspects of these projects are successful, the existing dominant patriarchy undermines the objectives of gender parity, reducing access and the influence of women on the outcomes of the projects.

These and other contributions to this volume all lead to similar conclusions. There are two essential aspects of policy and practice that is intended to sustain natural resources within a changing environment. One is that social science evidence on governance, policy implementation and related matters has to complement any natural science input. The other is that local communities must be an integral part of any programme development. Making the necessary changes to reduce environmental degradation, like all good deeds, begins at home.

The present volume is therefore a salutary reminder that the social sciences reveal issues crucial to sustaining natural resources. The challenges are not only physical and engineering ones. This volume therefore covers matters of great significance beyond the environmental sciences.

David Canter
Series Editor

Sustaining natural resources in a changing environment: evidence, policy and impact

Ruth Kattumuri

ABSTRACT
This article explores the global impacts of a changing environment on the sustainability of resources. In a global context characterised by continued population growth and accelerated urbanisation in emerging economies and the least developed regions of the world, pressures on environmental resources are intensifying. Extreme effects on ecosystems in both urban and rural communities are of enduring concern, as evidenced in water and food insecurity, and poor air quality. The author compares varying approaches to the collection and use of evidence, and the ways in which researchers may influence policy decisions and their implementation. Drawing on large and small-scale studies conducted in different regions of the world from a range of disciplinary perspectives, the article seeks to unravel the triangular relationship between research evidence, policy and impact, while paying attention to the tools used to assess impact on, and of, policy. In conclusion, the author considers how co-ordinated efforts by academics, public, private and third-sector practitioners across disciplines and national borders might produce stronger evidence and knowledge with which to inform decision-makers, empower citizens and achieve sustainable development, thereby supporting the needs of present and future generations.

Introduction

The availability of nutritious food, clean air and water are essential components in fostering ecosystems for a healthy life and improved well-being. In the early twentieth century, Mahatma Gandhi (1869–1948) articulated his belief that 'the earth provides enough to satisfy every man's need but not every man's greed'. He also stated that what we are doing to the forests of the world is a reflection of what we are doing to ourselves and to one another. The predictive power of his words has even greater relevance in the twenty-first century, as awareness has grown that the damage to the environment could be due to human factors, prompting action by researchers and policymakers to halt the degradation of natural resources, as exemplified in reports by the United Nations' Intergovernmental Panel on Climate Change (IPCC, 2014a), and as articulated in the Millennium and Sustainable Development Goals (SDGs).

Despite substantial improvements in the production of systematic information regarding changing environments and their impacts on natural resources, the use of available knowledge to inform policy continues to be hampered by low levels of confidence in much of the data and in many of the assumptions on which they are based (Leyshon, 2014). This article considers international and multidisciplinary research evidence of the possible impacts that changing biophysical and social environments can have on the sustainability of natural resources, and explores how such evidence is being used globally and nationally to influence policy decisions.

We begin by defining what we mean by a changing environment. We explore some of the policy relevant aspects of these changes that contribute to an understanding of the concept of sustainable development encompassing: population growth and urbanisation; public awareness of extreme weather events; attitudes to climate change predictions and perceptions of the power of governments to generate change.

We then examine the varying approaches adopted internationally and by different disciplines in the collection and use of evidence. The article reviews the ways in which researchers, professionals and practitioners from public, private and third sectors have sought to influence policy decisions and their implementation (Loorbach, 2010). It draws on large and small-scale studies conducted in different regions of the world, spanning North and South America, Africa, Asia and the Middle East, carried out by researchers from disciplines ranging from engineering and climatology, through the social sciences, to architecture, anthropology and philosophy. In seeking to unravel the triangular relationship between research evidence, policy and impact at global, national and local levels, attention is given to international targets for sustaining natural resources, and the tools used to achieve targets and assess research impacts. Our aim is to demonstrate the importance of developing multi-level thinking and planning for long-term sustainability, extending beyond short-term profit making and partisan politics, as argued by Sachs (2015) in his work on pathways to sustainable development, and by Jacobs and Mazzucato (2016) with reference to inter-sectoral co-production for sustainable and inclusive growth.

In conclusion, we consider how co-ordinated efforts by academics, public, private and third-sector practitioners, drawing on natural and social science methodologies, and different national perspectives, might become more effective in producing strong evidence and knowledge to inform decision-making, empower citizens and achieve sustainable development.

Global perspectives on changing environments

Much has been written about climate change and natural resources, and the causal relationship between them. In their Summary for Policymakers (SPM), the IPCC (2014a) unambiguously presents the current state of scientific thinking about the relationship between climate, its impact on natural and human systems and the ways in which governments should be tackling the problems facing humanity:

> SPM 1: Human influence on the climate system is clear .... Recent climate changes have had widespread impacts on human and natural systems;
>
> SPM 1.1: ... changes in climate have caused impacts on natural and human systems on all continents and across the oceans;

> SPM 1.3: Warming of the climate system is unequivocal, and since the 1950s, many of the observed changes are unprecedented over decades to millennia;
>
> SPM 4.5: Climate change is a threat to sustainable development. … there are many opportunities to link mitigation, adaptation and the pursuit of other societal objectives through integrated responses … . Successful implementation relies on relevant tools, suitable governance structures and enhanced capacity to respond … .

Within the confines of this short article, the intention is not to attempt to assess IPCC's scientific findings but to seek evidence of the extent to which their statements have, or have not, been acted on by policymakers in different parts of the world. In an analysis of 'critical issues in social science climate change research', Leyshon (2014), a geographer, addressed the ontological politics of climate change and the epistemological questions faced by social scientists tackling climate change. The present article draws selectively on a wider multidisciplinary literature to illustrate the relationship between research evidence, public perceptions and policymaking at global, national and local levels. Our interest in this section is in locating this relationship within the framework of the multifarious aspects of the changing environments in which natural resources are exploited and consumed, and in which public attitudes to the issues raised are formed and informed.

In a review of hierarchies of evidence in the context of the relationship between researchers and policymakers, political economists and social policy specialists have argued that 'no single source of evidence is likely to be determining or irrefutable' (Hantrais, Lenihan, & MacGregor, 2015, p. 108). Both across and within countries, sectors and socio-economic groups, not all evidence can be said to carry the same weight, or generate the same responses. These observations are particularly relevant in respect to climate change and sustainability of natural resources, where social science evidence is in competition for influence on public opinion and policymakers with scientific evidence and other factors such as political ideologies, long-held convictions and value systems. The wide range of evidence evoked in this section indicates the varied and conflicting interests of the research community studying the sustainability of natural resources in a changing environment. It also illustrates how different bodies of scientific evidence can be said to shape public perceptions and influence policy.

This section refers to case studies by researchers from the natural and social sciences and by practitioners working in different regions of the world. We consider the impacts on natural resources of both climate change and other contextual factors shaping the changing environment, including population growth and urbanisation. We examine how they affect public perceptions and attitudes, and exert additional pressure on policymakers in the public, private and third sectors, inciting them to develop and implement mitigation strategies.

The impact of population growth and urbanisation on natural resources

Underpinning current policy thinking at international level is the acknowledgement of the pressures on natural resources from population growth. The world population is currently around 7.5 billion and is estimated to reach 9.7 billion in 2050 (United Nations, 2017). The population in the least developed regions is predicted to continue growing, with an estimated 1.3 billion inhabitants to be added in Africa (doubling its current population), 0.9 billion in Asia and 0.2 billion in the rest of the world by 2050.

Around 54% of global population is currently urban according to United Nations' (2017) estimates. Evidence from urban geographers, sociologists and engineers shows how growing urbanisation is intensifying pressures on natural resources (Rout, 2018), creating extreme levels of air pollution (Srikanth & Nathan, 2018), the lack of clean water (Sansom, Hirst, & Kayaga, 2018), unequal access to energy (Brown et al., 2018), and continued food insecurity (Kenton & Singha, 2018) in both urban and rural communities (Zong & Cai, 2018). Much of the growth in mega cities (with a population of over 10 million) is predicted to be in less developed countries: 21 among 27 mega cities by 2025. The population in some mega cities already exceeds that of the estimates provided by the United Nations (2008) for that year: for example, the population of Tokyo is currently 37.8 million; the population of Delhi is estimated to have reached 25 million; and that of Shanghai, the most populated city in China, exceeds 24 million.

Major shifts in population from rural to urban environments, in conjunction with climate change, are recognised by international organisations to have intensified pressure on ecosystems (including crops, livestock, fisheries and forestry) and food security, resulting in an increase in the number of undernourished people since 2014 (FAO, IFAD, UNICEF, WFP & WHO, 2017). Evidence from a study carried out by environmental and ecophysiology scientists (Medek, Schwartz, & Myers, 2017) suggests that climate-related disruptions to agriculture will become more severe, with ever more serious implications for food supply, plant nutrients and human health. The authors conclude that 148.4 million of the world's population might be at risk of protein deficiency due to elevated atmospheric carbon dioxide concentrations (eCO_2) by 2050. The situation in sub-Saharan Africa, South Asia and especially India presents daunting challenges. They therefore recommend that eCO_2 effects on crop nutrient contents should be incorporated into future food security policies. An earth science study for the US National Aeronautics and Space Administration (NASA), based on evidence from global crop models collated by climate scientists, suggests, however, that eCO_2 levels could both 'help and hurt crops' (Reiny, 2016).

The US Global Change Research Program (USGCRP, 2015) was established by Presidential Initiative in 1989 and mandated by Congress in the Global Change Research Act of 1990 to 'assist the Nation and the world to understand, assess, predict, and respond to human-induced and natural processes of global change'. The USGCRP works with academics, the state, industry, and other groups on global change research to study climate change impacts on every type of natural resource from plants to animal species, and the alterations it brings to ecosystems. The Huron River Watershed Council (HRWC, 2016) was founded in 1965 in southeast Michigan and dedicated to river protection. The Council argues that increased risks of flooding, droughts and fires, and the severe degradation of natural resources are the result of reduced water levels, ground water recharge and wetland coverage, due to the warmer water and air temperatures.

The extent to which the American public believes such scientific evidence is revealed in an analysis by Marlon, Howe, Mildenberger, and Leiserowitz (2016) of the Yale Climate Opinion Maps. These authors used statistical modelling of national public opinion data gathered between 2008 and 2016 in the US. They found that: 70% of adults thought global warming was happening; 53% thought that it was due to human activities, while 32% attributed it to natural changes; 49% said that they believed evidence from scientists about global warming, whereas only 28% realised that no consensus existed; 71% said they somewhat/strongly trusted scientists, while 26% said that they did not. When Cook et al. (2013) analysed the evolution of the scientific consensus on anthropogenic global warming in peer-reviewed scientific literature, they found that, over time, the number of published papers rejecting the consensus on global warming had fallen to less than 3%. These findings suggest that, although members of the wider public in the US are rarely exposed to contradictory scientific evidence, they are not necessarily convinced by the views presented by professionals.

The European Commission's Joint Research Centre, Directorate for Sustainable Resource, analysed disaster records and high-resolution hazard and demographic projections for the period 1981–2010. Using a prognostic modelling framework through to the end of the twenty-first century, they examined human vulnerability in Europe due to extreme events: heat and cold waves, river and coastal floods, windstorms, droughts and wildfires. The evidence from their evaluation of spatial and temporal variations in intensity and frequency of greenhouse gas emissions demonstrated that, during the reference period, around 3000 Europeans lost their lives each year due to weather disasters (Forzieri, Cescatti, Silva, & Feyen, 2017, p. e202). The authors predicted that, in the absence of appropriate adaptation measures, the figure could reach 152,000 by 2071–2100, with about two-thirds of European population affected annually by weather-related disasters by 2100. They suggested that their results 'could aid in prioritisation of investments to address the unequal burden of effects of weather-related hazards and differences in adaptation capacities across Europe' (Forzieri et al., 2017, p. e207).

The role of technology in raising public awareness of changing environments

The rapid expansion of visual communication and social media has become an important component in evidence gathering and dissemination. By the end of the twenty-first century, technological developments had not only created a vast array of tools for assessing the impact of climate change on human and natural resources globally, they had also provided new instruments for raising and measuring public awareness of the impact of such changes on everyday life.

Technology has increasingly made it possible for citizens to engage directly by generating their own evidence through globally connected electronic media platforms, citizen journalism and other social media channels. Through print, audio and visual media, accounts of natural disasters and extreme weather events from across the world are disseminated on almost a daily basis, making the public aware of the adverse impacts on their everyday lives. Widely reported weather events in 2017 alone encompass: wild fires, droughts and crop damage in Spain and Italy (Henley, 2017); the heat wave in Asia, the Middle East and Europe (Masters, 2017); wind damage, hail, tornadoes and floods in the US (Dolce, 2017); mudslides in Sierra Leone (McAllister, 2017; Roy-Macaulay & Oyekanmi, 2017); and floods in South Asia (BBC, 2017).

Images such as those of the 2004 Indian Ocean Tsunami, the severe annual floods in Bangladesh, and the devastation caused by hurricanes Harvey, Irma, Jose, Katia and Maria in 2017, in rich and poor nations alike, play a crucial role in raising and maintaining public awareness of the impact of extreme weather events. Al Gore's film, 'The inconvenient truth' in 2006 and the Stern (2006) review on the economics of climate change, commissioned by Gordon Brown as UK Chancellor of the Exchequer in the Blair Government in the same year, were produced at a time when evidence based on events such as the Indian Ocean Tsunami had made the public receptive to new global policies. Fanny Armstrong's 2009 drama–documentary hybrid film 'The age of stupid', and its sequels, attracted crowd-funding, enabling it to achieve a global impact and to be credited with influencing the UK's energy policy (Puma, 2011).

In another example, a study in spring 2017 by the Pew Research Centre, a non-profit, non-partisan and non-advocacy US (fact) think tank, used global surveys and other social science instruments to inform the public about the issues, attitudes and trends shaping the world. The authors (Poushter & Manevich, 2017) reported that climate change is perceived as the greatest threat in 13 of the 38 countries surveyed, particularly in Latin America and Africa. Most of the other countries in the study ranked it as the second highest threat to national security. An earlier study by the Pew Research Centre suggested that people in countries with the highest per capita carbon emissions, such as the US, are less concerned about climate change and its potential implications than in Latin America and Africa where carbon emissions are much lower (Stokes, Wike, & Carle, 2015).

The same study found that 54% of respondents agreed with the statement that 'rich countries such as the US, Japan and Germany should do more than developing countries because they have produced most of the greenhouse gas emissions so far'; 38% were of the view that 'developing countries should do just as much as rich countries because they will produce most of the world's greenhouse gas emissions in the future'. The authors attributed differences in the level of concern to perceptions of the capability of governments to cope with the impacts. Across all the countries in the Pew study, 78% of respondents expected their governments to take greater responsibility for limiting greenhouse gas emissions as part of the international Paris agreement, while 67% accepted that the population will have to make lifestyle changes. Only 22% believed that technology would solve the problem without major changes in how people live (see also Canter, 2013; Gough, 2017).

In their examination of national public opinion in the US, Marlon et al. (2016) found strong support for policies to tackle climate change: 82% of respondents agreed that funding should be provided for research into renewable energies; 66% that utilities should be required to produce 20% of electricity from renewable sources; 75% that CO_2 should be regulated as a pollutant; and 69% that strict CO_2 limits should be set on existing coal-fired power plants.

Awareness of the major threats to natural resources and human security from the combined impact of population expansion, intensified urbanisation and extreme weather events associated with climate change has, undoubtedly, increased pressure for action at global level (Mitchell, Lowe, Wood, & Vellinga, 2006). As noted above (Marlon et al., 2016), a sizeable proportion of the US public remains sceptical about the causal relationship between climate change and human activities. However, public awareness and concern about the impact of changing environments on their livelihood, combined with their support for government intervention, could be said to lend legitimacy to mitigation and adaptation strategies at different levels of governance to reduce the impacts of climate change.

Collaborative international policy responses to evidence of changing environments

Several of the studies cited in the previous section report on large scale research carried out globally by international organisations. They show how sustainability of natural resources raises issues that engage researchers from multiple disciplines and sectors. Here, we consider further the opportunities provided by such engagement for collaboration across and within countries and sectors, particularly in the framework of international agreements and the assessment of internationally set targets. This section explores the contribution of researchers from a range of cultural and scientific backgrounds and disciplines, using a variety of methodological approaches, sometimes working collaboratively across the natural and social sciences, and usually in conjunction with governments and other stakeholders. These programmes and collaborations offer further insights into the relationship between evidence, policy and impact on the multiple aspects of sustainable environmental development at international, national and local levels.

Setting international targets

On 12 December 2015, the Paris Agreement on climate change was adopted by consensus by 195 of the 197 countries (excluding Syria and Nicaragua) subscribing to the United Nations Framework Convention on Climate Change. The Agreement was ratified by 160 countries and entered into force on 4 November 2016. Nicaragua subsequently indicated its intention to sign the Agreement in October 2017. The UN's Sustainable Development Goals adopted at a summit in September 2015 came into force on 1 January 2016. Evidence based on multiple bodies, systems, processes and 'advocacy coalition approaches' (Sabatier & Jenkins-Smith, 1993) was crucial in enabling these two agreements to be reached by countries jointly representing over 90% of world population. A key component in these global negotiations was the impact of the strong evidence presented regularly by heads of small island states concerning their vulnerability to global warming and extreme weather events (UNFCCC, 2005).

The United Nations' SDGs through to 2030 were designed to protect the planet from degradation by ensuring sustainable consumption, production and management of its natural resources, to promote urgent action on tackling climate change, and to end poverty. The growing body of evidence and knowledge about the vulnerability of the planet ensured that the majority of the seventeen SDGs focused directly, or indirectly, on sustaining natural resources in changing environments by improving the harmony between human beings and nature, and the well-being of global society. The most relevant goals in the context of the present article are those intended to:

SDG 2: end hunger, achieve food security and promote sustainable agriculture;

SDG 6: ensure availability and sustainable management of clean water;

SDG 7: ensure access to affordable, reliable, sustainable and modern energy for all;

SDG 8: promote sustained, inclusive and sustainable economic growth, full and productive employment and decent work for all;

SDG 11: develop resilient and sustainable cities;

SDG 12: promote responsible consumption and production;

SDG 15: protect, restore and promote air quality, sustainable use of terrestrial ecosystems, sustainable management of forests, combat desertification, and halt and reverse land degradation;

SDG 17: develop global partnerships for sustainable development.

(United Nations, 2015)

The SDGs have evolved through an inclusive process of intergovernmental negotiations between policymakers, through regular conferences and dialogues building on evidence and research, shared experiences, understanding, impact and lessons learned through 15 years of implementation of the Millennium Development Goals (MDGs). The UN SDGs set the targets and standards that need to be attained in full awareness that the levels of achievement would vary across different regions of the world, and be contingent on prevailing regimes and stages of development.

The case studies described here highlight the difficulties of achieving an effective relationship between the producers (researchers) and users of evidence (policymakers), even in instances where the evidence and its implications for policy in a local context appear to be strong and clear. A case study by Sansom et al. (2018), a team of UK and Ugandan civil engineers specialising in water engineering, considered how SDG 6 for universal piped water supplies could be reconciled with country realities in Sub-Saharan Africa. Their work demonstrates the importance of national medium-term target setting and reviewing and systematic planning approaches. They also identify the role played by careful global reporting of national performance against the SDG targets, portrayed as essential components for countries seeking to achieve these goals in the long-term. The research team worked closely with their stakeholders to achieve realistic targets in the medium term. Their article draws attention to the need to make adaptations to the SDGs when they are implemented in different environments to avoid demotivating countries with limited capacities and resources.

SDGs 2, 6, 7, 8, 12 and 15 were all relevant to a study by Srikanth and Nathan (2018) of opencast coal mining in India. The authors describe the potentially harmful environmental impact of surface opencast mines, due to loss of forests and habitats, disruption of biodiversity and of local communities, and associated damage to agriculture, water resources and local air quality. They identify the policy limitations in implementing the SDG principles due to inadequate guidelines, the lack of effective consulting mechanisms, and scarce local capacity and resources.

Although the SDGs were not explicitly targeted in their work, Zong and Cai (2018) applied international standard value measures in assessing the impact of urbanisation expansion on the cultivated land resources and food security in Chongqing, China, as sought in SDGs 2, 11 and 15. They describe the interactive relationship between researchers and policymakers in developing and implementing strategies to ensure national food security in response to evidence from scientific data about the impacts of the urbanisation process. Their account suggests how, in the Chinese context, government commissioned empirical studies are used to inform policy and to produce more efficient policymaking, in this case by researchers assisting governments in their efforts to protect the cultivated land.

Inter-sectoral collaboration

The United Nations MDGs involved public and private sectors, civil society and communities, and enabled them to work closely together to improve global development. In 2016, the Earth Security Group (ESG), a London-based inter-sectoral team from science, industry, finance and policy, carried out a study in collaboration with the Federation of Indian Chambers of Commerce and Industry and British High Commission in New Delhi. The study was designed to identify business innovations and policy partnerships enabling the insurance sector to play a more prominent role in building India's resilience to climate change in the energy and agriculture sectors (SDGs 2, 7). The ESG report, involving a forum of policymakers and business leaders, served as a catalyst for collaboration between insurers, industry and the Indian government. The Group drew on their strong networks within the policy community to communicate their findings to various ministers in India and to feed their evidence into discussions in the Indian Parliament in 2017.

Governments, international financial institutions and business leaders across the world are promoting insurance policies in the form of aid to developing countries with the aim outlined in SDG 17 of building global partnerships for sustainable development. In 2015, for example, based on available evidence about the impacts of climate change at global level, the G7 countries pledged to provide climate risk insurance for an additional 400 million people in the developing world by 2020, with Germany, Japan and the UK playing a lead role (Ralph & Aglionby, 2017).

Wiersma and Devine-Wright (2014) compared drivers and influences in projects concerned with the contribution of decentralised energy to the low carbon transition led by public, private and professional third-sector actors (SDG 7). They concluded that the substantial diversity within and between cases and across sectors is not sufficiently accounted for in academic projects interested in pursuing policy goals, again emphasising the importance of and in-depth understanding of a range of contextual factors determining the effectiveness of policy implementation.

Drawing on other internationally agreed standards and targets contributing to the attainment of SDG7, Oppong (2018) examined the dynamics of NGO participation in the Extractive Industries Transparency Initiative (EITI), an international auditing and multi-stakeholder oversight initiative adopted in Ghana in 2003. He highlights the negative effects of resource rents on state institutions, policy processes and economic output in resource-rich countries. His political economy analysis found that contestations within the NGO community around intervention processes inhibited meaningful implementation of evidence in resource governance. He concludes that, if the EITI is to avoid criticism for promoting transparency without improving accountability, it will need to achieve a closer fit with the strategic, normative and institutional values that different organisations, notably NGOs, attach to it in implementing countries (Figure 1).

When framed within wider national and international contexts, small-scale case studies are useful in shedding light on the ways in which social science evidence can be used both in policy and practice at the community level. Rout (2018) examines forest governance policies in India and their impact on the livelihood (in)security of forest communities, as

Figure 1. Artisanal miners in Ghana. © Keith Slack, Oxfam America. Reproduced with permission.

signalled in SDG 15. The evidence from his case studies shows how participatory models of governance based on bottom-up and rights-based approaches have sought to address environmental concerns. He argues that national policies, when implemented at the community level, can produce divergent and unwanted results if they do not take sufficient account of local contextual factors (Figure 2).

In another example of evidence produced at the community level, Kenton and Singha (2018) describe small-scale co-production action projects carried out by an international environmental design charity in the community development field. Their work in Venezuela and Palestine, two politically unstable countries, demonstrates how participatory methods can lead to greater community ownership and cohesion around shared concerns over access to healthy food and sustainable resource use in urban settings, reflecting the aims of SDG 2. Evidence from their studies, like that produced by Rout (2018), suggests how collaborative bottom-up processes can contribute to longer-term sustainable

Figure 2. People of the Malekudiya tribe discussing the development of their forest village. Photo courtesy of Anisha Sheth. Reproduced with permission.

policy development at the local level by encouraging effective governance of scarce resources in challenging environments.

Assessing research and policy impacts

The importance of scientific assessments of environmental impacts and the ability and willingness of governments to accept and act on their conclusions are demonstrated by Hochstetler (2018) in a comparison of the regulatory Environment Impact Assessment (EIA) frameworks in regional states in Latin America. She explores the arrangements for collecting, disseminating and assessing evidence about the environmental, social and economic impacts of projects including mining, hydroelectricity and wind farms. Her work shows that, although these reviews are most often conducted by an environmental agency involving some consultations with the local communities, current EIA frameworks have nevertheless become highly politicised. The tension created by producing socio-environmental evidence and economic policymaking may, she suggests, compromise the very results sought by the process.

The rationale and mechanics of impact and assessment of research on low carbon energy and international development are the focus of an article by a multidisciplinary team of geographers, civil engineers and anthropologists (Brown et al., 2018). The authors exploit networking and team analysis in considering how new policy initiatives, technological innovations and business models combining disciplinary, institutional, managerial, technological and methodological approaches can develop more effective evidence-based policy. They trace some of the debates around the conceptualisation of impact within different communities and by other stakeholders involved in research and innovation, as sought in SDG 7. Their analyses prompt them to argue for greater recognition and the attribution of specific resources to achieving longer-term network and capacity-building impacts beyond individual projects.

International and multidisciplinary perspectives

A substantial proportion of the reviews here on sustainability in changing environments derived from available literature from the US and/or is published in English-language media. However, the topic lends itself readily to international and multidisciplinary approaches that extend beyond the social sciences and English-speaking world. Several of the authors cited in this article have been associated with the UN or IPCC. These institutions rely on large-scale international teams using a variety of methodologies and tools across the sciences to carry out, assess and report their findings to the public and to policymakers. In addition, reference has been made to smaller scale evidence-based policy studies by researchers from Ghana, China and India, and to work with a policy dimension by native English speakers carried out in Brazil, Palestine, Uganda and Venezuela, illustrating wide-ranging interest in the topic.

Adopting a multidisciplinary approach, Brown et al. (2018) show how geography brings an appreciation of scale to questions of energy access. For example, in their focus on multi-level governance, they analyse how actors involved in governing the energy sector intersect at different scales. The authors acknowledge, however, that the community for low carbon energy research for development in the UK is fragmented and 'siloised'. The same observation could be applied to other countries, disciplines and areas of research. Opportunities are, Brown et al. suggest, being missed for linking disciplines and enhancing systemic complexity by cross-fertilising it with the innovations taking place in the private sector. Hence, multidisciplinary research will, they argue, only contribute to the challenges of securing universal access to a minimum acceptable amount of electricity, and the acceleration of low carbon transitions if it clearly responds directly to the demands and needs of communities and the variety of organisations (private, public or third sector) working towards those aims.

Despite the tendency in many countries for individual researchers to be siloised, several of the researchers cited in this article themselves have multidisciplinary and international backgrounds. For example, echoing the words of Mahatma Gandhi, cited in the introduction to this article, Gough (2017) brings to bear the multidisciplinary knowledge and methodologies of a social policy specialist in an analysis of *Heat, greed and human need: Climate change, capitalism and sustainable wellbeing.* Combining ethics and human need theory with political economy and climate science, he demonstrates how bridges can be built successfully between economy, ecology, social policy and politics. Oppong (2018), who is writing on extractive industries in Ghana, trained in political science, philosophy, international politics and development studies. He tackles his subject from the perspective of a political economist.

In their project on water targets in Uganda, Sansom et al. (2018) recognise that multidisciplinary work came about in at least two ways: firstly, by extending their areas of expertise, since two of the authors started their careers in engineering and then developed their expertise in water and sanitation management and institutional issues; secondly, through extensive interactions with experts over the years in other disciplines, including social scientists. The authors were thus able to contribute to research and advice on topics such as equity and community issues. Similarly, Srikanth and Nathan (2018) brought multidisciplinary and international perspectives to their case study in India. Nathan recorded three decades of experience in industry in the natural resources sector, and Srikanth's expertise extended to interdisciplinary research in sustainability and human development. Together, they contributed fresh insights by combining different but complementary perspectives from mining technology, mineral engineering management, policy and regulation, social consciousness and sustainability. Like Sansom et al. (2018), they worked closely with key stakeholders to explain the need for a unified regulator for surface coal mines to overcome current policy and regulatory gaps. The Chinese geographers Zong and Cai (2018) examined the urbanisation process in a city located in the western part of China from a spatial perspective. They also worked closely with policymakers to reveal the different impacts of urbanisation on land resources.

Kenton and Singha (2018) combined the knowledge and expertise of an architect and a development practitioner with the experience of directing an international environmental design charity working in the field of disaster relief amid political upheaval in low- and middle-income countries. Their projects show how architects, engineers, development activists and designers are collaborating globally to safeguard environmental and cultural sustainability (Singha, 2012). The authors used methodologies that are locally relevant, empowering, and replicable to carry out community-led bottom-up projects, thereby offering a way forward for improved policymaking and efficient governance of resources at the local level in a landscape of rapid change and scarce resources.

As argued by Hochstetler (2018), 'politics' is often blamed for environmental problems, and the most powerful economic and political actors frequently reach their position by indiscriminately using and degrading natural resources. Many solutions require collective authority as well, often from the state, necessitating the construction of new, more sustainable, political coalitions. Political scientists are, therefore, well placed to contribute to the body of knowledge about sustainable development. In examining how Brazilian impact assessment works in practice, Hochstetler (2018) acknowledges that natural scientists and engineers provided much of the information that she used. Her analysis required a social scientific analysis to assess social and economic changes, illustrating the key role played by social scientists at all stages of international and multidisciplinary projects.

Conclusion

An important aim of this article was to compare and contrast international and multidisciplinary perspectives on the development of policy in response to evidence about the impact of environmental change on the sustainability of natural resources. To this end, the article has explored both the institutions involved in generating and mediating evidence, and the different methodologies employed in collecting and assessing evidence, informing policy, and contributing to governance.

In the field of sustainability of natural resources, the challenge of generating unbiased evidence with high levels of confidence requires a strong commitment and collaborations among researchers from a multiplicity of disciplines and sectors. The scale of the impacts and vulnerabilities due to climate change has encouraged physical and social scientists to work together in an unprecedented manner to gain a better understanding of the challenges it presents. IPCC (2014b) recognised the importance of understanding decision-making contexts in determining the type and scale of climate change risks based on physical climate science and impacts, adaptation and vulnerabilities assessment. The case studies reported in this article indicate how the effectiveness of implementation is often determined by a range of socio-economic and political contextual factors affecting the ability of researchers and policymakers to communicate and cooperate in setting, agreeing and achieving international targets.

While acknowledging the conceptual and contextual problems of working across national, disciplinary and sectoral boundaries (British Academy, 2016; Hantrais, 2009; Lunn & Ruane, 2013), this article has considered whether and how co-ordinated efforts by academics, public, private and third-sector practitioners can achieve reciprocal impacts. Drawing on natural and social science methodologies, and different national perspectives, our analysis suggests that collaborations across and between sectors can, if properly managed, lead to the more effective generation and transfer of knowledge that informs decision-making and empowers citizens while seeking to achieve sustainable development.

Notes on contributor

Ruth Kattumuri, FAcSS is Founder and Co-Director of the LSE India Observatory and an Associate at the Grantham Research Institute on Climate Change and the Environment. She is involved in transdisciplinary evidence-based research, policy influence and advice concerning sustainable growth, inclusion and development, primarily in the context of the Global South. Her recent work has been on adaptation and mitigation for climate change, covering the interface with food, water and energy security; low carbon energy transitions; urban water resource management and governance; and low carbon urban growth and development.

References

BBC. (2017). *Deadly South Asia floods affect 16m people. (Broadcast 18 August)*. Retrieved from http://www.bbc.co.uk/news/world-asia-40975232

British Academy. (2016). *Crossing paths: Interdisciplinary institutions, careers, education and applications*. British Academy Report. London: British Academy. Retrieved from https://www.britac.ac.uk/sites/default/files/Crossing%20Paths%20-%20Full%20Report.pdf

Brown, E., Campbell, B., Cloke, J., Cross, J., Long, S. T., Turner, B., & Wray, A. (2018). Low carbon energy and international development: From research impact to policymaking. *Contemporary Social Science, 13*(1), 112–127. doi:10.1080/21582041.2017.1417627

Canter, D. (2013). Why do we leave it so late? Response to environmental threat and the rules of place. *Journal of Earth Science & Climatic Change, e5*, 169. doi:10.4172/2157-7617.1000169

Cook, J., Nuccitelli, D., Green, S. A., Richardson, M., Winkler, B., Painting, R., … Skuce, A. (2013). Quantifying the consensus on anthropogenic global warming in the scientific literature. *Environmental Research Letters, 8*(2), 024024. Retrieved from http://iopscience.iop.org/article/10.1088/1748-9326/8/2/024024/meta

Dolce, C. (2017). *2017 off to destructive start: Severe weather reports tally 5000+, more than double the average. (Blog 10 April)*. Retrieved from https://weather.com/storms/tornado/news/severe-weather-hail-tornado-wind-damage-2017-mid-april

Earth Security Group (ESG). (2016). *Indian insurance and sustainable development*. Report by ESG, FICCI and British High Commission New Delhi. Retrieved from https://earthsecuritygroup.com/wp-content/uploads/2016/06/ESG.IndiaInsurance.pdf

FAO, IFAD, UNICEF, WFP & WHO. (2017). *The state of food security and nutrition in the world 2017: Building resilience for peace and food security*. Rome: FAO. Retrieved from http://www.fao.org/state-of-food-security-nutrition/en/

Forzieri, G., Cescatti, A., Silva, F. B., & Feyen, L. (2017). Increasing risk over time of weather-related hazards to the European population: A data-driven prognostic study. *The Lancet Planetary Health Journal, 1*(5), e200–e208. Retrieved from http://www.thelancet.com/pdfs/journals/lanplh/PIIS2542-5196(17)30082-7.pdf

Gough, I. (2017). *Heat, greed and human need: Climate change, capitalism and sustainable wellbeing*. Cheltenham: Elgar.

Hantrais, L. (2009). *International comparative research: Theory, methods and practice*. Basingstoke: Palgrave Macmillan.

Hantrais, L., Lenihan, A. T., & MacGregor, S. (2015). Evidence-based policy: Exploring international and interdisciplinary insights. *Contemporary Social Science, 10*(2), 101–113. doi:10.1080/21582041.2015.1061687

Henley, J. (2017, August 18). Extreme heat warnings issued in Europe as temperatures pass 40C. *The Guardian*. Retrieved from https://www.theguardian.com/world/2017/aug/04/extreme-heat-warnings-issued-europe-temperatures-pass-40c

Hochstetler, K. (2018). Environmental impact assessment: Evidence-based policymaking in Brazil and Latin America. *Contemporary Social Science, 13*(1), 100–111. doi:10.1080/21582041.2017.1393556

Huron River Watershed Council (HRWC). (2016). *Climate change impacts on natural resources*. Retrieved from http://www.hrwc.org/wp-content/uploads/2016/11/HRWC-Natural-Resources.pdf

Intergovernmental Panel on Climate Change (IPCC). (2014a). *Climate change 2014–impacts, adaptation and vulnerability: Part B: Regional aspects: Working group II contribution to the IPCC fifth assessment report*. New York, NY: Cambridge University Press.

IPCC. (2014b). *Climate change 2014 synthesis report summary for policymakers*. Retrieved from https://www.ipcc.ch/pdf/assessment-report/ar5/syr/AR5_SYR_FINAL_SPM.pdf

Jacobs, M., & Mazzucato, M. (Eds.). (2016). *Rethinking capitalism: Economics and policy for sustainable and inclusive growth*. Oxford: Wiley–Blackwell.

Kenton, N., & Singha, S. (2018). Community empowerment in changing environments: Creating value towards food security. *Contemporary Social Science, 13*(1), 85–99. doi:10.1080/21582041.2017.1417630

Leyshon, C. (2014). Critical issues in social science climate change research. *Contemporary Social Science, 9*(4), 359–373. doi:10.1080/21582041.2014.974890

Loorbach, D. (2010). Transition management for sustainable development: A prescriptive, complexity-based governance framework. *Governance, 23*(1), 161–183. doi:10.1111/j.1468-0491.2009.01471.x

Lunn, P., & Ruane, F. (Eds.). (2013). *Using evidence to inform policy*. Dublin: Gill & Macmillan.

Marlon, J., Howe, P., Mildenberger, M., & Leiserowitz, A. (2016). *Yale climate opinion maps – U.S. 2016*. Retrieved from http://climatecommunication.yale.edu/visualizations-data/ycom-us-2016/

Masters, J. (2017). *Historic heat wave sweeps Asia, the Middle East and Europe. (Blog 6 June)*. Retrieved from https://www.wunderground.com/cat6/historic-heat-wave-sweeps-asia-middle-east-and-europe

McAllister, E. (2017). *Cities across Africa face threat of landslides like Sierra Leone. (Blog 18 August)*. Retrieved from http://uk.reuters.com/article/uk-leone-mudslide-africa-idUKKCN1AY119

Medek, D. E., Schwartz, J., & Myers, S. S. (2017). Estimated effects of future atmospheric CO2 concentrations on protein intake and the risk of protein deficiency by country and region. *Environmental Health Perspectives, 125*(8), 8. doi:10.1289/EHP41

Mitchell, J. F. B., Lowe, J., Wood, R. A., & Vellinga, M. (2006). Extreme events due to human-induced climate change. *Philosophical Transactions of the Royal Society A: Mathematical, Physical and Engineering Sciences, 364*, 2117–2133. doi:10.1098/rsta.2006.1816

Oppong, N. (2018). Negotiating transparency: NGOs and the contentious politics of the extractive industries transparency initiative in Ghana. *Contemporary Social Science, 13*(1), 58–71. doi:10.1080/21582041.2017.1394483

Poushter, J., & Manevich, D. (2017). *Globally, people point to ISIS and climate change as leading security threats: Concern about cyberattacks, world economy also widespread. Pew Research Centre Report*. Retrieved from http://assets.pewresearch.org/wp-content/uploads/sites/2/2017/07/31101043/Pew-Research-Center_2017.07.13_Global-Threats_Full-Report.pdf

Puma. (2011). *The age of stupid. Puma Creative Impact Award*. Retrieved from http://www.docimpactaward.org/files/ageofstupid.pdf

Ralph, O., & Aglionby, J. (2017). *The push for faster aid to developing countries. Financial Times (15 August)*. Retrieved from https://www.ft.com/content/cdd65f6e-7dc7-11e7-9108-edda0bcbc928

Reiny, S. (2016). *NASA Study: Rising carbon dioxide levels will help and hurt crops (3 May)*. Retrieved from https://www.nasa.gov/feature/goddard/2016/nasa-study-rising-carbon-dioxide-levels-will-help-and-hurt-crops

Rout, S. (2018). Gendered participation in community forest governance in India. *Contemporary Social Science, 13*(1), 72–84. doi:10.1080/21582041.2017.1393555

Roy-Macaulay, C., & Oyekanmi, L. (2017, August 15). *Authorities warn of more mudslides in Sierra Leone as toll hits 350 dead with hundreds more missing*. Retrieved from http://time.com/4906112/sierra-leone-mudslide-landslide-freetown/

Sabatier, P. A., & Jenkins-Smith, H. (1993). *Policy change and learning: An advocacy coalition approach*. Boulder, CO: Westview Press.

Sachs, J. (2015). *The age of sustainable development*. New York: Columbia University Press.

Sansom, K., Hirst, D., & Kayaga, S. (2018). International water targets and national realities in Sub-Saharan Africa: The case of Uganda. *Contemporary Social Science, 13*(1), 17–29. doi:10.1080/21582041.2017.1393557

Singha, S. (2012). *Architecture for rapid change and scarce resources*. London: Routledge.

Srikanth, R., & Nathan, H. S. K. (2018). Towards sustainable development: Planning surface coal mine closures in India. *Contemporary Social Science, 13*(1), 30–43. doi:10.1080/21582041.2017.1394484

Stern, N. (2006). *The economics of climate change: The Stern review*. Cambridge, MA: National Bureau of Economic Research.

Stokes, B., Wike, R., & Carle, J. (2015). *Global concern about climate change, broad support for limiting emissions*. Pew Research Centre. Retrieved from http://www.pewglobal.org/2015/11/05/global-concern-about-climate-change-broad-support-for-limiting-emissions/

UNFCCC. (2005). *Climate change: Small island developing states*. Bonn: Climate Change Secretariat. Retrieved from http://unfccc.int/resource/docs/publications/cc_sids.pdf

United Nations. (2008). *World urbanization prospects, 2007 revision*. New York, NY: United Nations.

United Nations. (2015). *Transforming our world: The 2030 agenda for sustainable development. A/RES/70/1*. Retrieved from https://sustainabledevelopment.un.org/content/documents/21252030%20Agenda%20for%20Sustainable%20Development%20web.pdf

United Nations. (2017). *World population prospects: The 2017 revision*. New York, NY: United Nations.

US Global Change Research Program (USGCRP). (2015). *Water resources, land use & land cover, ecosystems & biodiversity, adaptation*. Report sponsored by US Department of Agriculture Forest Service. Retrieved from http://www.globalchange.gov/sites/globalchange/files/ASIWG_Synthesis_4.28.15_final.pdf

Wiersma, B., & Devine-Wright, P. (2014). Decentralising energy: Comparing the drivers and influencers of projects led by public, private, community and third sector actors. *Contemporary Social Science*, *9*(4), 456–470. doi:10.1080/21582041.2014.981757

Zong, H., & Cai, B. (2018). The impact of urbanisation on land resources in Chongqing, China, 1997–2015. *Contemporary Social Science*, *13*(1), 44–57. doi:10.1080/21582041.2017.1417628

International water targets and national realities in Sub-Saharan Africa: the case of Uganda

Kevin Sansom, David Hirst and Sam Kayaga

ABSTRACT
This article considers how to reconcile ambitious UN Sustainable Development Goals (SDGs) for universal piped water supplies with developing country realities in Sub-Saharan Africa (SSA). A concise process for effective reviews of medium-term national targets is proposed and is applied in an analysis of the current provision of piped water to households and shared community facilities in urban and rural settings in Uganda. Different disciplinary perspectives are adopted to review trends, the performance of key stakeholders and their scope for achieving new targets. Only about 5% of households have piped water supplies on their premises in rural areas in SSA. To achieve the SDG target of 100% coverage will, therefore, take a long time and requires continued support for the sustainability of community water facilities as a priority. The SDGs offer sensible long-term aims, but national medium-term target setting and reviewing remain critical and require realistic and systematic planning approaches, as well as careful global reporting of national performance against SDG targets. The authors argue that balanced incentives are needed to encourage stakeholders to strive for realistic targets in the medium term, without demotivating countries with limited capacities and resources.

Introduction

As the transition between the Millennium Development Goals (MDGs) and the Sustainable Development Goals (SDGs) (2016–2030) is made, it is important to consider key strengths and weaknesses of the processes and consequences of the MDGs. The United Nations (UN) recognised that the MDG agenda promoted concrete actions to address human development shortfalls, provided a common framework and improved coordination for development actors. However, the MDGs resulted in the setting of rather rigid national policy agendas, following international benchmarks rather than local conditions, often ignoring the complexities of the development process (UN, 2012, pp. 6, 7). In addition, even where water MDG targets were met by 2015, there are doubts about the extent to which the new facilities will be sustainably managed, particularly in rural areas.

The SDGs were developed following an exhaustive process of debates and negotiations (UNDESA, 2015, p. 1). Research evidence shows that, when the water, sanitation and hygiene (WASH) SDGs are eventually achieved, they will bring considerable and wide-ranging benefits (Hutton & Chase, 2016, p. 20). Meanwhile, more needs to be known about the challenges that governments are facing in developing sustainable water services, and how SDG targets are going to be met.

During the MDG period up to 2015, households were counted as having access to potentially safe water sources if they used 'improved' water sources for individual households or shared sources that include hand-pumps or stand-posts. In Uganda, compliance with the 'improved water source' criteria included households within 1000 m (rural), whereas the SDG for drinking water (target 6.1) relates to water supply on the premises of all households (UNWater, 2016, pp. 12, 13), thereby considerably raising the bar for target achievement.

Only 5% of households in rural Africa currently have piped water on their premises; most people rely on shared point water sources such as hand-pumps and public stand-posts (WHO & UNICEF, 2015, p. 77). The proportion rises to 10% when other on-plot options such as boreholes are included. In urban Africa, where the economies of scale for piped water are more effective, the percentage of households with piped water on premises has only risen from 44% to 46% since 2000 (WHO & UNICEF, 2017, p. 105). The SDGs are, therefore, overly ambitious in expecting a further 90% of rural households to gain new on-plot water supplies by 2030.

The clear risk is that, if governments and international donors focus on SDG targets for on-plot water solutions, they may neglect poorer rural areas where only shared water facilities, such as hand-pumps, are currently feasible. A key SDG objective is to 'leave no-one behind', which is a desirable equity objective, but many rural communities are poor, remote or vulnerable, and will struggle to develop and manage effective piped on-plot water systems, without extensive assistance from government and donors over a long period.

Uganda has been selected as a case study because of its good record in water sector performance and reforms (Water and Sanitation Program, World Bank, 2011, p. 15; WHO & UNICEF, 2015, p. 75). In Uganda, the Ministry of Finance, Planning and Economic Development leads the process in developing medium-term plans that covers key sectors. For example, in 2017, the National Development Plan II was agreed for 2015–2016 to 2019–2020 (Ministry of Finance Planning and Economic Development, 2017). The Plan is based on budget framework papers with, for instance, the Ministry of Water and Environment (MWE), and incorporates trends in performance, planning and budget forecasts. This process is enhanced through comprehensive annual water sector performance reporting and joint sector reviews. However, given the major shift in emphasis in the SDGs, as explained above, future medium-term plans and targets need to take into account critical analyses of key stakeholders' capacities and approaches. This issue is examined here and is also incorporated into a suggested framework for national water sector medium-term target setting.

Assessments for water targets

Water sector institutions in each country are required to balance their funding priorities and plans for meeting ambitious SDG targets and providing basic services to poor and hitherto unserved populations in a realistic and iterative way, bearing in mind absorptive capacity on a sustainable basis. The broader question addressed in this article is how to reconcile ambitious SDGs for universal piped or on-plot water supplies with developing country realities. This issue is examined considering the following research questions:

(1) How has the Ugandan water sector performed in terms of service levels, equity and management aspects, and what are key reasons for the trends identified?
(2) How have key types of rural and urban water sector stakeholders performed to date in Uganda, and to what extent are they likely to be able to achieve proposed SDG targets?
(3) What are the implications of the SDG targets, considering challenges and opportunities for rural and urban water services in Uganda?
(4) How best can national governments develop and work towards realistic medium-term WASH targets, while addressing the longer term SDGs?

These questions are consistent within the broader framework for reviewing medium-term development targets at national level, as suggested in Figure 1. The research reported in this article is based on recent field research on the management of rural water supplies in Uganda, utilising previous urban water research conducted by two of the authors. A review of quantitative and qualitative data has also been conducted using the available literature on the subject and includes extensive grey literature, academic articles and UN publications.

Reed, Fraser, Morse, and Dougill (2006, p. 414) propose a rather complex 'adaptive learning process for sustainability indicator development and application'. Our aim is to suggest how the proposed targets can be periodically reviewed in terms of service levels in a specific national context. A concise overview process, such as the one shown in Figure 1, can be used by key national stakeholders in planning for improved WASH service provision. Some aspects shown in Figure 1 are not considered in detail, such as

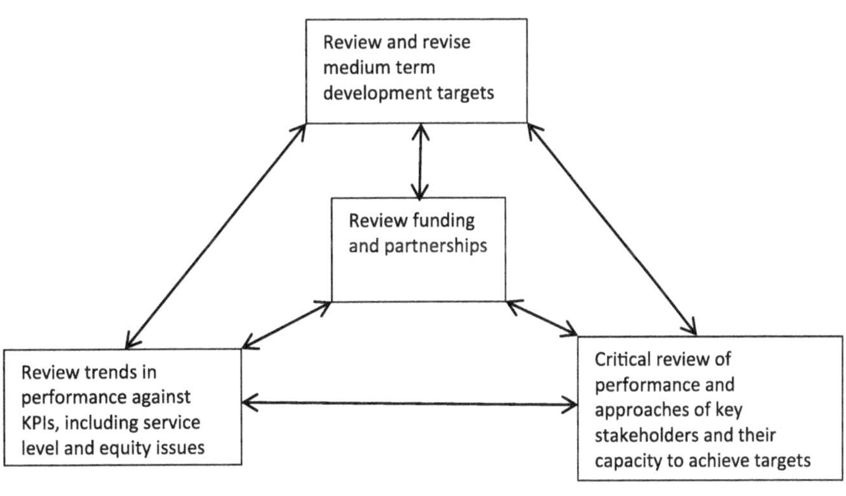

Figure 1. Reviewing medium-term national WASH development targets. Source: The authors.
Note: KPI: key performance indicators.

funding and partnerships. Without systematic reviews of medium-term targets, national governments may find that the gap between the ambitious SDG targets and what they can actually achieve on an annual basis leads to poorly focused plans.

SDG target 6.1, states: 'By 2030, achieve universal and equitable access to safe and affordable drinking water for all' (WHO, 2017, pp. 12, 13). Table 1 summarises what this means in terms of level of services for both the SDG target and the more basic MDGs.

The MDGs mainly relate to shared water sources, whereas, in order to meet the SDG target 6.1 threshold for a 'safely managed' service, the improved source must meet three conditions:

- the source should be located on premises (within the dwelling, yard or plot);
- water should be available when needed;
- the water supplied should be free from faecal and priority chemical contamination.

If any of the three conditions is not met, but the improved source is within 30 minutes of the home, it will continue to be categorised as a 'basic' service. 'Improved' sources are those that are potentially capable of delivering safe water by nature of their design and construction (WHO, 2017, pp. 12–13). Sustainability is a key concern in the rural sector, as can be seen from the Uganda case study. The WHO acknowledges that individual countries will need to consider how to balance the support for extending access (MDGs) and to improve service levels (SDGs) (WHO, 2017, p. 13), which is difficult to achieve in situations where sector funding is limited.

SDG Target 6b) states: 'Support and strengthen the participation of local communities in improving water and sanitation management'. This is necessary to ensure that solutions decided upon are suitable for specific socioeconomic contexts and to encourage local ownership of the solutions when implemented to ensure sustainability over time (UNWater, 2016, p. 25).

Community involvement is an important part of such stakeholder participation. However, a growing body of evidence (Lockwood, Smits, Schouten, & Moriarty, 2010, p. 4) reveals the limitations and lack of sustainability of community management of water supplies in rural areas. In many African countries, sustained efforts have been made to increase the involvement of the private sector, particularly in undertaking repairs of water facilities.

Equity of services 'implies progressive reduction and elimination of inequalities between population subgroups' (WHO, 2017, p. 11). Jarman (2016, p. 110) concluded that 'age, ethnicity and gender are still fundamental variables that together and separately contribute to differentiated social outcomes'. This observation clearly has relevance in

Table 1. The new JMP ladder for household drinking water services in the SDG era.

MDG/SDG	Service ladder	Progressive realisation
SDG 6.1	Safely managed drinking water	Improved source located on premises, available when needed, and free from faecal and priority chemical contamination
MDG continuity	Basic water	Improved source within 30 minutes round-trip collection time
	Limited water	Improved source over 30 minutes round-trip collection time
	Unimproved water	Unimproved source that does not protect against contamination
	No service	Surface water

Source: WHO (2017) pp. 12, 13.

rural Africa where 'Women and girls are responsible for water collection in 8 out of 10 households with water off their premises' (WHO, 2017, p. 30). The new SDG basic service level 'Improved source within 30 minutes round-trip collection time' and SDG target 6.1 shown in Table 1 should contribute to a significant reduction in the burdens on women and children. Ndesamburo, Flynn, and French (2012, p. 453) highlight 'the need for equity and inclusion approaches to better understand and meet the water, sanitation, and hygiene (WASH) needs of women and girls, as well as other groups who are often excluded'.

Equity of service provision is taken here to mean the extent of use of basic and unimproved service levels and by wealth quintile in Uganda. National sector performance reporting in Uganda (MWE, 2016, pp. 26, 128) also regularly reviews equity issues by area-based variations in the distribution of community water points and by gender-based indicators, namely 'the percentage of water committees or boards with women holding key positions'.

The Ugandan case study

Multiple data sources were used in the case study as part of a triangulation process. For the rural water field research, 15 semi-structured interviews, 5 focus group discussions and observations were conducted on Hand Pump Mechanic Associations (HPMAs) in 5 districts, together with key informant interviews at national level. The selection of these research locations in Uganda was carefully considered against several criteria including: availability of data, presence and years of HPMA operation and presence of alternative operation and maintenance (O&M) models. The urban water research was based on a thorough review of the many government utility and consultancy reports in Uganda, building on previous research by the authors and the academic literature.

The three main parts of this section assess current progress and prospects for achieving the SDG water targets in Uganda, as set out in Figure 1. Firstly, we review piped water supplies to households, which are mostly found in urban areas; secondly, an assessment is made of shared community water facilities in rural areas and thirdly, we review progress in achieving equity in the water sector.

Piped water services and stakeholder capacities in Uganda

The percentage of households in urban Uganda who have piped water supplies on their premises has steadily increased from 7% in 1990 to 23% in 2015 (WHO & UNICEF, 2015, p. 75), despite a high annual population growth of over 3%, a figure that is supported by recent household surveys and the MWE 2016 sector performance report, but is higher than the coverage reported in the recently launched JMP report (WHO & UNICEF, 2017).

The National Water and Sewerage Corporation (NWSC), a public utility, is performing well in managing water services in the larger Ugandan towns against a number of indicators and ratios. For example, the number of customer pipe connections has increased from 152,138 in 2006 (MWE, 2016, p. 23) to 472,193 in 2016 (MWE, 2016, p. 45). NWSC is also maintaining good service levels, as is demonstrated by the customer satisfaction index at 88% (MWE, 2016, p. 45). Other aspects include carefully managing customer relations, employing competent staff, while generating funds for investment. NWSC have also made good progress in enabling urban poor households to access piped water services (MWE, 2016, p. 45; NWSC, 2015).

Despite these achievements, a number of challenges remain. The average non-revenue water (NRW) in large towns, that is the percentage of water produced and not paid for excluding households who are not connected to the piped water system, is 31.6% (NWSC, 2015, p. 8), compared to a desirable level of 20% (McIntosh, 2003, p. 62). However, NWSC compares well with other utilities in Sub-Saharan Africa (SSA) in terms of trends in performance: it was assessed as the second best performing national utility in 2013 (Van den Berg & Danilenko, 2017, p. 47).

Currently, NWSC provides water services in 218 towns (see https://www.nwsc.co.ug/). The remaining towns are under the responsibility of the MWE through various Water Authorities and/or private operators. Of the towns under MWE, 60 do not have piped water supply (MWE, 2016, p. 32).

Private operators manage water supplies in a few small towns in Uganda as part of formal management contracts (MWE, 2016, p. 32), while local governments oversee these contracts. Private operators have managed small town water services with some success, although providing incentives for new capital investments and paying for expansion of water services, with such short-term management contracts, have proved problematic (Hirn, 2013, p. IV). Increasingly, NWSC has taken on the management of more small-town water supplies, but such towns have higher levels of NRW and poorer service levels (MWE, 2016, p. 53). NWSC staff reported in 2017 that they are effectively using a 'clustered approach' to managing small-town water services, whereby managers in larger towns are supporting staff managing water services in nearby smaller towns. However, those small-town water systems are usually cross-subsidised by the larger town offices. Overall, Uganda has good capabilities for managing larger piped water systems effectively.

In considering the prospects for developing piped water services in rural areas, it is relevant to examine what is currently happening with piped supplies in these areas. 'Typically a user elected Water and Sanitation Committee (WSC) is responsible for the management of village piped water supply services and a local individual is hired for the operation of the scheme' (Hirn, 2013, p. 6).

To achieve universal household piped water services, a key question emerges: Is there sufficient capacity and incentive to manage small rural piped water systems, where the economies of scale are considerably less, and affordability is more of an issue? Private operators or capable individuals would seem to have the potential to manage small rural systems in conjunction with local committees, provided that capital investments are organised centrally. There are also plans for NWSC to provide piped water services to rural growth centres (see https://www.nwsc.co.ug/). However, current sector funding levels suggest that a large expansion of piped water systems in rural areas will be difficult to achieve on a sustainable basis (WHO & UN-Water, 2017, p. IX). An African Ministers Council on Water (AMCOW) Country Overview for 2015 and beyond for Uganda (Water and Sanitation Program, 2011, p. 2) highlighted that Uganda, with its strong institutions, has witnessed remarkable improvements in access, but the low investment anticipated for the future may see these gains eroded.

Most people undoubtedly prefer the convenience of piped water supplies on their premises and are willing to pay for these services. This is evident from the 97% customer bill payment or collection efficiency for NWSC piped connections (MWE, 2016, p. 47). However, many poorer households who reside in rural areas find it difficult to pay for on-plot piped water services, as is demonstrated by the fact that only 1% of rural households have piped connections (WHO & UNICEF, 2015, p. 75). The figure rises to 4% when other on-plot options such as boreholes are added (WHO & UNICEF, 2017, p. 75). Rural households are, therefore, likely to continue to rely on shared water sources for many years to come.

Community water point services and stakeholder capacities in rural Uganda

Hand-pumps are widely used for the rural poor in Africa as a cheaper and safer means of community drinking water sources, compared to many alternatives, but keeping hand-pumps functioning in the medium to longer term using community management continues to present major challenges. About 35% of hand-pumps in Africa are not functioning at any one time (Rural Water Supply Network, 2009, p. 1), even though most countries in Africa have adopted the prevailing paradigm of community-based management systems (CBMS) for achieving functional and reliable rural water supplies.

In Uganda, improvement in the functionality of rural water supplies increased to 88% in 2014 but fell to 86% in 2015, which is still below the expected targets set by the Government of Uganda of 90% functionality. The fall in functionality is attributed to non-repair of broken-down facilities (MWE, 2016, p. 3). Furthermore, although functionality levels are better than the global picture, only 53% of supplies are fully functional in that they provide an adequate and reliable yield throughout the day (MWE, 2011, p. vi), suggesting that the CBMS model is not delivering in its current form.

The prevailing school of thought in the literature is that the community management model has clear limitations and has failed to deliver expected levels of sustainability. Sustainable, community-based managed RWS continues to remain an elusive goal. Greater efforts are needed to test and evaluate alternative models for managing hand-pump water supplies (Foster, 2013, pp. 12037–12045).

A sector-wide approach (SWAp) in the Ugandan rural water sector has enabled 'the building of impressive government and donor alignment, harmonization, and coordination' (WSP, World Bank, 2011, p. 15). A number of challenges remain, according to the MWE, as shown in Table 2.

O&M is a key area in the search for sustainable RWS. Due to the limited success of the CBMS, a number of complementary and alternative models have emerged, an example being greater use of the local private sector, as part of public–private partnerships (PPPs). A number of hand-pump centred models have been developed in Uganda. We consider HPMAs in more detail here, as this approach has been scaled up in districts across Uganda.

The emergence of 'Community Management Plus' indicates the need to provide more professionalised services, such as the HPMA. The HPMA model has been incorporated in government policy and is being implemented throughout the country in at least 85 out of 108 districts surveyed, who now have registered HPMAs (MWE & SNV, 2015, p. 5).

An early example of the development of an HPMA in 2005/2006 was when WaterAid, an international UK-based charity, created an association of hand-pump mechanics in Katakwi district (Uganda). The main objectives were to improve the coordination of the work of the HPMs who undertake repairs, and enhance the credibility and quality

Table 2. Main challenges to CBMS operation and maintenance systems and causes.

Main challenges	Causes
(1) More water users lose trust in WSC	Embezzlement of water fees by water and sanitation committee (WSC) members. Also possible Illegal intervention by politicians
(2) Water users come to expect that water is free	Water users whose distrust has grown refuse to pay water fees and WSC cannot collect water fees on a regular basis
(3) Caretaker retires, and no one is left to take over	WSC, which cannot collect water fees on a regular basis, refuse to remunerate caretakers
(4) Hand pump parts gradually suffer wear and tear and pump discharge begins to decrease	The absence of caretakers means that the hand pump cannot be managed and preventive maintenance cannot be done
(5) Hand pumps breaks and cannot be fixed promptly because WSC have not saved for repairs	Hand pumps cannot be promptly repaired when they are broken because WSC with no regular collection of water fees have not saved for maintenance expenses. WSC lack sufficient funds for maintenance because hand pumps break frequently
(6) Hand pumps remain out of use for long periods of time and water users' trust in WSCs plummets dramatically	Water users' trust in water supply projects deteriorates due to hand pumps breaking frequently or being out of use frequently or for long periods
(7) Governments have little financial power and cannot respond to all requests for repairs from WSCs	Repairs become expensive because there is no choice but to use private entities to perform them. Very few conditional subsidies (DWSSCG) are granted
(8) Hand pumps cannot be managed and preventive maintenance cannot be done because water fees are not being collected regularly	Caretakers to manage hand pumps cannot be hired because WSC is unable to collect water fees on a regular basis. WSC do not understand the importance of preventive maintenance because broken hand pumps can be used again if they are repaired with government support

Source: National annual water and sanitation sector performance report (MWE, 2016, p. 25).

assurance of the work done by the HPMs. The association supervises the work of the HPMs, supports them with transport and reports to the districts. The districts provide the associations with office space and stationary, and organise training for all HPMs, while the water facilities are managed by the village water committees. The district council also monitors the HPMS and regulates charges for maintenance and repairs, to avoid exploitation of the communities. The districts concerned have seen an impressive increase in functionality and reduction of downtime of the hand pumps.

The field study of HPMAs by the authors examined the effectiveness and capacity of the HPMA model in five districts and was analysed according to four categories. HPMAs scored highest in technical capacity, poorly in financial management, with average ratings in management/leadership and operational/business capacity. These findings suggest that HPMAs have continued with the technical repair work that they have been carrying out for several years, but have been less inclined to adopt some of the more managerial and financial aspects of the HPMA model, possibly due to the lack of incentives put in place to expand their services. A key issue is the reluctance of communities to pay the HPMs for preventative maintenance, although it costs less in the long run, compared to paying for corrective maintenance when a more serious problem occurs.

The fact that 80% of the case study HPMAs were able to perform complex rehabilitation tasks, such as fishing pipes out of the boreholes, is indicative of how far they have come by learning from one another. Variations in scores between HPMAs can be attributed to several factors including longevity of the association, strength of the enabling environment and availability of business opportunities (field interviews).

The HPMA approach has established some foundations that can potentially improve CMBS in the coming years. However, the model currently lacks a sufficiently supportive enabling environment if it is to become fully functional. Moreover, the focus on corrective repairs at the expense of preventative maintenance raises doubts over its sustainability.

Another approach being piloted is the 'Whave' PPP model, which focuses on preventative maintenance and hence has better prospects of being more cost effective than purely corrective maintenance options. The Whave PPP model has been implemented in several districts in Uganda and involves working through local water service providers (WSPs) who are individuals contracted by the Water Operator (in this case Whave Solutions) to provide O&M services to a specific number of water points. The model is performance-based, focusing on preventive maintenance rather than repair. The WSPs earn money from a functioning water source rather than a broken-down water source.

The model has had significant impact on functionality rates, which, at 99% is very encouraging. Nonetheless, some design issues with the Whave model need to be resolved if the model is to be introduced on a wider scale. The model is subsidised with Whave contributing about 40% and the community 60% of the total service cost. These models are promising and are being considered for future scaling up by the MWE and development partners. Meanwhile, the challenge of providing sustainable rural water services remains, despite extensive efforts to address these various issues as part of the SWAp for water in Uganda. Government, donors and NGOs have previously focused rural water sector support on community-managed water sources. Who will provide sufficient resources and support for developing piped water systems in a sustainable manner as part of the drive to achieve SDG 6.1 is an outstanding question.

Equity in service provision

The SDG 6 targets for water supply emphasise 'universal access', which means for everyone. This is consistent with the overarching ambition of the 2030 Agenda to 'end poverty in all its forms' and 'leave no one behind' (WHO, 2017, p. 10). Analysis of the Ugandan data shows that only 4% of the urban population have unimproved (potentially unsafe) main sources of water, whereas the percentage of the population using unimproved water supplies in rural areas was 24% in 2015. Despite this significant reduction, many people are still facing public health risks due to the unsafe water that they drink (WHO & UNICEF JMP, 2015). Improving the water sources for those 24% in rural areas with community shared water sources is clearly the priority, rather than providing more expensive piped water services.

Disaggregated data for per income quintile in 2012 for water supplies in Uganda show significant variations in equity. In urban areas, 60% in the richest quintile, compared to only 5% in the poorest quintile, had piped water on their premises (WHO & UNICEF, 2015). In rural areas, only the richest quintile displayed a significant percentage of people with piped water on premises.

In the Uganda annual sector performance report (MWE, 2016, p. 26), equity of provision between areas is defined in terms of access to water points measured by calculating the standard deviation between the average numbers of persons per improved water point at sub-county level. The variation has changed from 243 persons per water point in 2007/2008 to 142 in 2015/2016, and is explained in part due to data entry problems. The trend is nevertheless encouraging. In the same annual report, the authors track gender equity against the Key Performance Indicator of per cent of Water User committees/

Water Boards with women holding key positions. Performance against this target has risen to 86% in 2016. Gender mainstreaming initiatives are thus found to have resulted in capacity building in 24 districts (MWE, 2016, p. 128).

Overall improvements have been made in the various dimensions of equity in recent years. However, large gaps remain between urban and rural areas, in both Uganda and SSA. In particular, wide disparities are found in the provision of household piped water supplies, which are mainly confined to richer households, suggesting that provision of piped water supplies to poorer households (SDG 6.1) is likely to be hampered in the future due to affordability and poverty issues, particularly if sustainable services are to be provided.

Discussion and conclusions

Compared to many other African countries, Uganda has good capabilities for managing larger piped water systems effectively within both the national water utility and private sectors. However, limited incentives exist for managing small-scale rural piped water systems, where economies of scale are considerably less, and affordability is more of an issue in rural areas. A substantial increase in funding and support would be needed for new water infrastructure and sustainable management if the SDG targets are to be achieved. A key focus would also be needed on equity to provide and sustain piped water sources for poor communities.

In SSA, 68% of the total population have 'improved water sources' compared to 79% for Uganda (WHO & UNICEF JMP, 2015, pp. 75, 77). Uganda is also performing better than most African countries in terms of 'transitioning to country-led service delivery pathways for water supply' according to AMCOW (2011). The challenges in improving water services in most SSA countries are greater than in Uganda, the more so since 'Aid commitments for water and sanitation to SSA have declined from US$ 3.8 billion to US$ 1.7 billion from 2012 to 2015' (WHO & UN-Water, 2017, p. IX).

According to WHO & UN-Water (2017, p. IX), National WASH budgets are increasing as countries prepare to take on board the SDGs. Yet discrepancies remain between global aspirations and national realities. More than 80 per cent of countries report insufficient financing to meet national WASH targets, let alone the higher levels of service that are the focus of the SDGs.

The SDG emphasis on providing piped or on-plot water supplies for all households is recognised as a very ambitious long-term goal. For this reason, substantive support from donors and development partners for achieving fully functioning sustainable community water points will still be required for many years, particularly for poor rural areas. Due to the hitherto limited success of community-based management of shared water sources, a number of alternative 'Community Management Plus' approaches have emerged that support the local private sector in providing key inputs to enable sustainable outcomes.

Innovative institutional mechanisms and public–private–people partnerships will be needed to provide both sustainable piped water supplies and better management of shared community water supplies in rural areas of SSA. Such mechanisms are best developed based on a careful analysis of the lessons learnt and good practice in Uganda and elsewhere.

Our analysis of the Ugandan case suggests that innovative institutional mechanisms and public–private–people partnerships will be needed to provide both sustainable piped water supplies and better management of shared community water supplies in rural areas of Sub-Saharan Africa. Such mechanisms can best be developed based on a careful analysis of the lessons learnt and good practice in Uganda and elsewhere.

The water SDGs provide sensible long-term aims for the types of services that most people prefer, but tracking of progress against the SDG's targets needs to be managed carefully. During the MDG period, the JMP regularly publicised which countries were on- or off-track against water and sanitation targets. However, this approach failed to account adequately for the considerable progress made by countries with low initial levels of human development, especially in Africa (UN, 2012, p. 7).

Publication of performance tracking against targets can provide incentives for policy-makers, where the targets are feasible, but reliance on over-optimistic targets can have a demotivating effect among in-country stakeholders and create uncertainties around priorities. According to the economist Coelho (2010, p. 406): 'Unrealistic optimism creates distortions which may be the most important source of efficiency loss in the economic system. When designing policy, governments often ignore this fact'.

How best then can we reconcile ambitious SDG targets with national realities? International performance reporting of progress against SDGs for piped water services in rural areas requires careful management, with less emphasis on simplistic on- and off-track reporting. The percentage increases in households with piped connections or similar indicators for each country could be a suitable indicator for constructive performance tracking.

Our analysis suggests that achievable medium-term country targets need to be given greater emphasis. A concise and robust method for medium-term target setting and review is critical, and is best achieved at national level, using realistic and systematic planning approaches. Balanced incentives seem to be necessary to encourage stakeholders to strive for realistic targets in the medium term, without demotivating countries with limited capacities and resources.

Disclosure statement

No potential conflict of interest was reported by the authors.

Notes on contributors

Kevin Sansom is a lecturer and programme manager at WEDC (Water, Engineering and Development Centre), in the School of Architecture, Building and Civil Engineering at Loughborough University, UK. He has undertaken strategic research, consultancy, and capacity development projects in the management of water and sanitation in Africa and Asia, for both the urban and rural sectors. He has extensive experience in programme monitoring and evaluation, sector reviews, institutional development, collaboration with non-state providers and water utilities management. He has had key roles on projects in more than 18 countries, over 25 years, working with development agencies, governments, NGOs and other stakeholders.

David Hirst completed his studies for an MSc in Water and Environmental Management, at the Water Engineering Development Centre, in the School of Architecture, Building and Civil Engineering of Loughborough University. He is a development practitioner with over ten years of project and staff management experience in water and sanitation, agricultural livelihoods and food security programmes in East and West Africa, including eight years working in Uganda. His main research area of interest is the sustainability of rural water systems with a particular focus on improved models of operation and maintenance.

Sam Kayaga is a chartered Water/Sanitation Engineer and Senior Lecturer in the School of Architecture, Building and Civil Engineering, Loughborough University. Prior to joining Loughborough University's Water, Engineering and Development Centre (WEDC) in 2002, he worked for eighteen years for the National Water and Sewerage Corporation (NWSC), Uganda's urban water utility. At Loughborough, he is involved in research, teaching and advisory services for sustainable water and sanitation services in middle- and low-income countries. His main areas of research are sustainable urban water and sanitation management and institutional capacity, with a special focus on extending services to the currently unserved.

References

African Ministers Council on Water. (2011). *AMCOW country status overviews – regional synthesis report. Pathways to progress: Transitioning to country-led service delivery pathways to meet Africa's water supply and sanitation targets*. Washington, DC: The World Bank. Retrieved from http://www.wsp.org/wsp/content/pathways-progress-status-water-and-sanitation-africa

Coelho, P. M. (2010). Unrealistic optimism: Still a neglected trait. *Journal of Business and Psychology, 25*(3), 397–408. doi:10.1007/s10869-009-9146-9

Foster, T. (2013). Predictors of sustainability for community-managed hand-pumps in Sub-Saharan Africa: Evidence from Liberia, Sierra Leone, and Uganda. *Journal of Environmental Science & Technology, 47*(21), 12037–12046. doi:10.1021/es402086n

Hirn, M. (2013). *Private sector participation in the Ugandan water sector: A review of 10 years of private management of small town water systems* (Water and Sanitation Program: Working Paper). Nairobi: World Bank. Retrieved from https://www.wsp.org/sites/wsp.org/files/publications/Private-Sector-Participation-in-the-Ugandan-Water-Sector-Review.pdf

Hutton, G., & Chase, C. (2016). The knowledge base for achieving the sustainable development goals targets on water supply, sanitation and hygiene. *International Journal of Environmental Research and Public Health, 13*(6), 1–35. E536. doi:10.3390/ijerph13060536

Jarman, J. (2016). Social inequality and its consequences in the twenty-first century. *Contemporary Social Science, 11*(2–3), 103–112. doi:10.1080/21582041.2016.1254277

Lockwood, H., Smits, S., Schouten, T., & Moriarty, P. (2010). *Providing sustainable water services at scale* (International Symposium on Rural Water Services: Background Paper). Retrieved from https://www.ircwash.org/sites/default/files/Lockwood-2010-Providing.pdf

McIntosh, A. (2003). *Asian water supplies: Reaching the urban poor*. Mandaluyong: Asian Development Bank (ADB).

Ministry of Finance Planning and Economic Development, Uganda. (2017). National Development Plan II has just been agreed for 2015–16 to 2019–20. Retrieved from http://www.gou.go.ug/ministry/ministry-finance-planning-and-economic-development

Ministry of Water and Environment. (2011). *Assessment of the effectiveness of community-based maintenance systems for rural water supply facilities*. Kampala: Directorate of Water Development, Ministry of Water and Environment, the Republic of Uganda.

Ministry of Water and Environment, Uganda. (2016). *Water and environment sector performance report*. Kampala: Government of Uganda.

Ministry of Water and Environment, Uganda & SNV. (2015). *National Assessment of District Hand Pump Mechanics Associations in Uganda*. Draft Report II, Kampala.

National Water and Sewerage Corporation. (2015). *NWSC Corporate Plan 2015–18*. Retrieved from: https://www.nwsc.co.ug/files/corprateplan/NWSC_CORPORATE_PLAN_2015-2018

Ndesamburo, J., Flynn, E., & French, S. (2012). Looking through an equity and inclusion lens in Tanzania: The experience of WaterAid. *Gender & Development, 20*(3), 453–465. doi:10.1080/13552074.2012.731809

Reed, M. S., Fraser, E. D. G., Morse, S., & Dougill, A. J. (2006). An adaptive learning process for developing and applying sustainability indicators with local communities. *Ecological Economics Journal, 59*(4), 406–418. doi:10.1016/j.ecolecon.2005.11.008

Rural Water Supply Network. (2009). *Hand-pump data in selected countries in Sub-Saharan Africa*. Retrieved from http://www.rural-water-supply.net/_ressources/documents/default/203.pdf

UNDESA. (2015). Sustainability development goals – interactive timeline of post-2015 and post-Rio +20 processes? Retrieved from http://www.sustainabledevelopment2015.org/index.php/timeline

United Nations. (2012). *Review of the contributions of the MDG Agenda to foster development: Lessons for the post-2015 UN Development Agenda*. New York. Retrieved from: http://www.un.org/en/development/desa/policy/untaskteam_undf/group_a_mdg_assessment.pdf

UNWater. (2016). Integrated monitoring guide for SDG 6 targets and global indicators. Retrieved from: http://www.unwater.org/app/uploads/2017/03/SDG-6-targets-and-global-indicators_2016-07-19.pdf

Van den Berg, C., & Danilenko, A. (2017). *Performance of water utilities in Africa*. Washington, DC: World Bank.

Water and Sanitation Program, World Bank. (2011). *An AMCOW country status overview – water supply and sanitation in Uganda – turning finance into services for 2015 and beyond*. Retrieved from http://documents.worldbank.org/curated/en/212801468111263063/Water-supply-and-sanitation-in-Uganda-turning-finance-into-services-for-2015-and-beyond

WHO. (2017). *Safely managed drinking water - thematic report on drinking water*. Geneva: World Health Organization.

WHO & UNICEF. (2015). *Estimates of use of water and sanitation facilities and by wealth quintile for Uganda*. Retrieved from https://www.wssinfo.org/data-estimates/

WHO & UNICEF. (2017). *Progress on drinking water, sanitation and hygiene: 2017 update and SDG baselines*. Geneva: World Health Organization.

WHO & UNICEF JMP. (2015). *Progress on sanitation and drinking water update, Joint Monitoring Programme*. Geneva: World Health Organization.

WHO & UN-Water. (2017). *Financing universal water, sanitation and hygiene under the sustainable development goals* (Report on Global Analysis and Assessment of Sanitation and Drinking-Water (GLAAS)). Geneva: World Health Organization.

Towards sustainable development: planning surface coal mine closures in India

Raman Srikanth and Hippu Salk Kristle Nathan

ABSTRACT
Coal is the major source of India's electricity today, accounting for 59 per cent of its electricity generation capacity and 75 per cent of the electrical energy generated. Given that 63 per cent of the power generation capacity added in the Twelfth Five-Year Plan (2012–2017) was coal-based, coal is set to remain the most staple source of electricity for India in the foreseeable future. Of the coal produced in India, more than 90 per cent is dispatched from surface or opencast mines, with potentially harmful effects on the environment, due to loss of forests and habitats, disruption of biodiversity and of local communities, and associated damage to agriculture, water resources and local air quality. Acknowledging these adverse environmental impacts, the Government of India has mandated the restoration of mining areas post mine closure to create a 'self-sustaining ecosystem', while optimising the use of mined-out land for the benefit of local communities. Within this context, the article reviews India's surface coal mine closure policies, regulatory regimes and operating practices with reference to best practices for reclamation and restoration in selected major coal-producing countries. The article identifies the shortcomings in India's policies and suggests strategies and measures to remedy them.

Introduction

India is the third largest electricity (power) generator in the world after China and the US (BP, 2017). The major primary energy sources that contribute to India's power generation are coal, natural gas, oil, nuclear, hydro and different renewable sources like wind, solar, bio-power and small hydro power.

Coal dominates India's power mix with a generation capacity of 59 per cent (Central Electricity Authority [CEA], 2017). Due to the relatively higher utilisation or plant-load factors of coal-based power plants, coal contributes an even larger proportion of electricity generation, with contributions exceeding 60 per cent since 1985; and more than 70 per cent since 2012 (World Development Indicators, 2017).

Given that 63 per cent of the power generation capacity added in the Twelfth Five-Year Plan (2012–2017) is coal-based (CEA, 2017), coal will continue to be the mainstay of India's electricity in the coming decades. The National Electricity Plan (CEA, 2016) has estimated that, even in the most optimistic scenario of increased generation capacity of renewable sources, power generation from coal could increase. According to International Energy Agency (2016) estimates for 2040, coal will still constitute 41 per cent of installed power generation capacity and fuel 55 per cent of electricity generation in India.

Approximately three-fourths of the coal used in India is produced domestically. With a global share of 9.3 per cent, India is the second largest coal producer in the world (BP, 2017). The power sector is the largest consumer of coal in India, consuming more than three-fourths of the coal produced (Coal Controller's Organisation [CCO], 2017). The dependency on domestic coal in India is due to its relatively higher endowment compared to other fossil fuels. India has 8.3 per cent of world's proven coal reserves, but only 0.3 and 0.7 per cent of the world's oil and natural gas reserves, respectively (BP, 2017).

Coal mining in India has a long history, in 1774, Summer & Heatley applied to the East India Company to extract coal from the Raniganj coalfield in Bengal (CCO, 2017). However, until the setting up of a rail link between Howrah and Raniganj in 1855, coal mining in India received little attention due to the inferior quality of Indian coal compared to that produced in UK (CCO, 2017). From its humble origins, the coal industry in India grew rapidly, particularly after India's independence in 1947 (CCO, 2017; Guha, 2002; Ministry of Information and Broadcasting, 1990, Chapter 7; Sharma, 2000).

One of the key features of India's coal sector is that most of the coal is obtained through surface mining. India has 512 coal and lignite mines in operation, consisting of 252 underground mines, 229 surface mines and 31 mines with both underground and surface mining operations (CCO, 2017). Surface mines account for the largest share of coal produced. In 2015/2016, 593 million tonnes (93 per cent) of the 639 million tons of coal produced in India were from surface mines (CCO, 2017). The preference for surface mining in India is due to its higher extraction rate, lower cost of production and better safety record when compared to underground mining (Satyanarayana, 2012).

Although surface mining is safer, more economical and far more productive, it has adverse environmental consequences. Major environmental issues include erosion, loss of forest and biodiversity, contamination of soil, air and water (both surface and ground water), changes to the flows and discharges of rivers and streams, and wind direction, damage to building and structures, global warming and other related consequences (Central Pollution Control Board, 2011).

Mining activity can have some positive impacts as well. It has an economic dividend in terms of infrastructure development of the local area and provision of employment and business opportunities for local people. The coal industry is one of India's largest employers. For example, Coal India Ltd. (CIL) provides direct employment to approximately 314,000 persons (a further 99,000 persons are employed by contractors) in addition to generating hundreds of thousands of indirect jobs in eight states of India (Ministry of Coal [MOC], 2017).

However, surface coal mining requires the acquisition of large areas of land for mining, waste dumping and other ancillary activities. The issues raised by land acquisition for surface coal mines are exacerbated in India given the large population and high population density. At times, like other mega infrastructure development projects, surface coal mining leads to involuntary displacement of people. Dissatisfaction with compensation benefits in return for loss of land and the appreciation of land value post-development create a sense of public apathy and resentment towards many energy projects (Raj, Rethinaraj, & Nathan, 2016).

The Government of India has acknowledged the adverse environmental consequences of surface coal mining on air, water and land in the adjoining areas (Ministry of Environment, Forest and Climate Change, 2016). This adverse environmental impact will become more critical in future, since the total volume of excavation required to produce coal is continuing to increase. More effective policies, technologies and practices are therefore needed to counter these threats to the environment.

During its long history, the coal industry in India operated without any statutory guidelines for surface coal mine closure until Mine Closure Guidelines were formulated for the first time by the Ministry of Coal (MOC) in 2009 (MOC, 2013). Several coal mines were closed or abandoned prior to the issue of these guidelines and were not subject to statutory obligations on the part of either the mine owner or the government to complete their closure and rehabilitation. Following the adoption of the mine closure guidelines in 2009, approximately 500 operating coal mines were identified by CIL and Singareni Collieries Company Ltd. (SCCL) for closure (MOC, 2017). By 2015, 198 coal mines were closed or abandoned in coalfields currently mined by CIL (MOC, 2016).

In 2013, the Government of India revised its guidelines for the closure of coal mines: stating that mining areas must be restored to create a 'self-sustaining ecosystem' (MOC, 2013). Additionally, the Mines and Minerals Development & Regulation (MMDR) Act of 2015 (Ministry of Law and Justice [MLJ], 2015) explicitly requires restoration and reclamation activities to make optimal use of mined-out land for the benefit of the local communities. However, these goals are yet to be spelt out in terms of actionable policies and practices. Against this backdrop and using documentary sources, personal observation and interviews with key stakeholders, the authors review India's surface coal mine closure policies, regulatory regimes and operating practices with reference to best practices for reclamation and restoration to identify gaps, and indicate key corrective measures and strategies that could be adopted to achieve India's stated goals.

The framework for mine closures in India

The overarching principle of mine closure is premised on the tenet of sustainable development. The Brundtland Commission report (World Commission on Environment and Development, 1987, p. 16), which popularised this concept, defines sustainable development as 'development that meets the needs of the present without compromising the ability of future generations to meet their own needs'. With reference to coal mining, these needs are different across generations. For the present generation, the need is for a reliable and affordable fuel resource, whereas future generations need open green spaces, water resources and the facilitation of economic development including development of alternative energy resource (for instance, the restored land can be used for a solar or wind farm). The self-sustained ecosystem enshrined in India's mine closure guidelines can be interpreted to imply that the restored land must provide a revamped ecosystem and economic services for the benefit of future generations, particularly those who reside in the surrounding areas.

The governing principle of restoration of mined-out lands is drawn from India's legal system, based on English Common Law, and includes the public trust doctrine as part of its jurisprudence: the State is the trustee of all natural resources, which are by nature meant for use and enjoyment of the public at large. Hence, the State as a trustee is under a legal duty to protect natural resources (Supreme Court of India [SCI], 1996). In this context, the Supreme Court of India (SCI) has held the 'precautionary principle' and the 'polluter pays principle' to be essential features of sustainable development (1996).

According to the SCI the precautionary principle means that government and statutory authorities must anticipate, prevent and tackle the causes of environmental degradation. Where there are threats of serious and irreversible damage, the lack of scientific certainty should not be used as a reason for postponing measures to prevent environmental degradation (SCI, 1996). This principle also puts the 'onus of proof' on the developer or actor to show that their actions are environmentally benign (SCI, 1996). The polluter pays principle implies that absolute liability for harm to the environment extends not only to compensation for the victims of pollution, but also to the cost of repairing environmental degradation (SCI, 1996).

The overarching principle of mine closure is also reflected in the UN's Sustainable Development Goals (SDGs), which India adopted in 2015. SDG 8 is to promote sustained, inclusive and sustainable economic growth, full and productive employment and decent work for all; and SDG 12 is to ensure sustainable consumption and production patterns (United Nations Development Programme, 2015). These two goals are in line with the guiding principles for restoring the mined-out land to a self-sustaining ecosystem for the benefit of the local communities.

Mine closure is not a one-off activity which begins only after a mine is closed and ends when the mine is reclaimed and restored in accordance with the approved Mine Closure Plan. Rather, it is a continuous process spanning the entire life of the mining activity as depicted in Figure 1.

The Government of India acknowledges the continuity of the process, and clearly demarcates the different stages in the mine closure process, involving progressive or concurrent mine closure, and final mine closure (MOC, 2013). Progressive mine closure includes various land-use activities to be carried out continuously and sequentially during the entire period of the mining operations, whereas the final mine closure activities start towards the end-of-mine life and may carry on even after mining operations are discontinued until the mining area is restored to create a self-sustained ecosystem (International Council on Mining and Metals [ICMM], 2008; MOC, 2013). The coal mine owner is required to prepare Progressive Mine Closure Plans at intervals of five years from the beginning of the mining operations, while the details, including updated cost estimates of mine closure activities, must be included in the Final Mine Closure Plan which shall be submitted to MOC at least five years before the mine closure deadline (2013).

Although final mine closure is normally scheduled when all the minable reserves within the delineated mining lease are exhausted, due to some practical difficulties, such as geotechnical complications, economic infeasibility or social protests, the closure may take place much earlier (ICMM, 2008). The real possibility of such events occurring in the mining industry justifies continuous planning for mine closure at all stages. Decisions taken during the mine planning and development stage, and even earlier, during the exploration stage can have profound effects on the ultimate closure plan, its cost and the resulting landscape performance (ICMM, 2008).

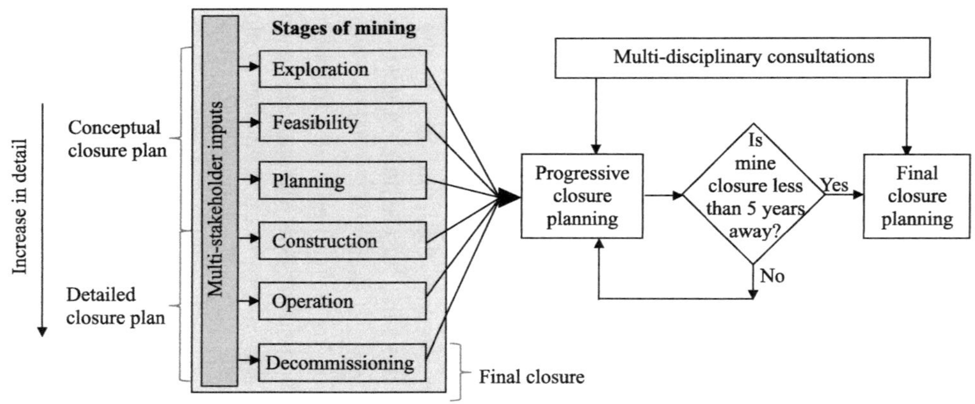

Figure 1. Framework for mine closure planning.

Note: Adapted from (ICMM, 2008; MOC, 2013).

The process of closure planning evolves during the life of the mine. It is initially conceptual but progressively becomes more detailed as the mine nears decommissioning and closure to ensure successful land reclamation (ICMM, 2008). The ICMM (2008) has identified a three-step evolution process in its toolkit (see Figure 1). The conceptual plan communicates the outcomes and goals, whereas the detailed plan includes milestones and standards, detailed methodologies for achieving them, as well as monitoring and validation processes to ensure that these planned milestones are achieved. The conceptual plan has an active life of 3–5 years and may extend to the planning and construction stage with few revisions, whereas the detailed plan has an active life of 5–30 years and is revised continually. The final mine closure plan has a life span depending on post-closure responsibilities.

The overarching aim of a Mine Closure Plan is to allow productive and sustainable post-mining use of the site that is acceptable to all key stakeholders. The challenge is to align the interests of the three stakeholders: to reduce the environmental liability of the 'mine owner'; to maximise benefit to 'local communities' and to comply with the sustainable development agenda of the 'government'. Mine closure thus needs to satisfy multi-pronged objectives: to conserve valuable attributes of pre-mining land; to protect public health and safety; to alleviate environmental damage and to provide socio-economic stability (Ontario Mining Act, 2005).

The process necessitates a multi-disciplinary expertise dedicated to understanding the geological, social, biological, forestry, hydrological, hydrogeological and environmental aspects of mine closure and rehabilitation. Throughout the mine's life, closure strategies and plans need regular review to ensure that they are appropriate and address the major issues for mine closure, while remaining aligned with stakeholders' expectations and regulatory requirements. Hence, multi-stakeholder and multi-disciplinary consultations become an essential part of a mine closure framework (see Figure 1). Since these consultations are key to the success of any mine closure process, they should ideally continue throughout a mine's life from planning to post-mining closure. If such consultations are conducted effectively, they enable the mine owner to build trust and mutual respect at all stages of the process (ICMM, 2008).

The mine closure process ultimately determines what is left behind as a legacy or benefit for future generations. From the mine owner's point of view, closure of a mine can also be an opportunity to establish an enduring positive legacy, which not only results in lasting benefits to local communities even after the owner has left the site (ICMM, 2008), but also provides an opportunity to ensure a smoother process in obtaining permits for future projects by earning the goodwill of key stakeholders. By contrast, if the mine closure does not meet the required standards and leaves some key stakeholders dissatisfied, the owner risks his reputation, which may lead to delays or even failure in obtaining social and environmental approvals for ongoing and future projects.

Worldwide best practices compared

Some of the major coal-producing countries have demonstrated productive use of mined-out land both for commercial purpose and for the betterment of local communities. While the US and Australia, with coal production levels of 607 and 493 million tons in 2016, respectively, rank behind India with its 692 million tons of coal production (BP, 2017), both countries have well-established regulatory mechanisms governing surface coal mines. The interface between mining and environmental regulators in the Republic of South Africa which has a coal production of 251 million tons in 2016 (BP, 2017) is also of interest, because, at the time of writing, it was restructuring its regulatory processes for the mining sector.

Invariably, the countries quoted for their best practices consider mine closure in the broader context of sustainable development. In the preamble to the 1977 Surface Mining Control and Reclamation Act (SMCRA), the US Congress explicitly acknowledged that many surface mining operations reduce the utility of the land for all purposes – commercial, industrial, residential, recreational, agricultural and forestry – by causing erosion and landslides, contributing to floods, polluting air and water, destroying fish and wildlife habitats, creating hazards dangerous to life and property by degrading the quality of life in local communities (SMCRA, 1977). Recognition of the multiple dimensions and levels of the adverse effects of surface coal mining resulted in comprehensive Mine Closure Plans in the US.

Similarly, in Australia, several multi-disciplinary – geology, hydrology, climate, hydrogeology and limnology – studies have been conducted in pit lakes, along with long-term monitoring of water balance and water quality (Zhao, McCullough, & Lund, 1999). The regulators governing surface coal mines in both the US and Australia are qualified in multiple disciplines, equipped with suitable resources, and have acquired the specialised experience required to regulate surface coal mines that have multiple impacts on air, water, land, biodiversity and socio-economic structure of the mine and its adjoining areas (Commonwealth of Australia, 2016; Office of Surface Mining and Reclamation [OSMRE], 2017).

The US and Australia have well-established standards against which the approach and progress to mine closure adopted by the coal mine owners are evaluated. These standards are not one-size-fits-all solutions; and allow innovative solutions when required, exemplified in the manual published by the Federal Government in Australia on a 'Leading Practice Sustainable Development Program for the Mining Industry' (Commonwealth of Australia, 2016). Further, several closed mine sites in Australia have been adapted for tourism, agricultural and ecological uses, demonstrating innovative solutions.

The US mandates that coal-mined land should be returned to its approximate original contours in most cases; exceptions are agreed when the community requests different contours to achieve more beneficial post-mining land uses (SMCRA, 1977). During their 40 years of experience in implementing the SMCRA, the Office of Surface Mining Reclamation and Enforcement (OSMRE) and different State regulators overseeing surface coal mines in accordance with OSMRE standards have developed a variety of regulations, guidelines and procedures to improve the collection of environmental resource information, predict impacts, and plan for mitigation and reclamation (OSMRE 2017). In another noteworthy example of innovation, the lignite mining pits of Saxony and Brandenburg in former East Germany were converted into a series of 26 lakes with a well-planned rejuvenation programme ('East Germany's old mines', 2016).

Mine closure presents a complex mixture of environmental, social, economic and development issues, requiring governments to play a pivotal role in the entire process. In the countries cited for their best practices, governments go to great lengths to ensure that industry adequately recognises and prepares for eventual closure over the life of the mine, and that the Mine Closure Plan is carried out to the satisfaction of the communities involved, other major stakeholders, and government at all levels.

In the US, the intervention of the government is justified on the basis that the burden of surface mining adversely affects commerce and the public welfare by counteracting governmental programmes and efforts to conserve soil, water and other natural resources (SMCRA, 1977). In countries with a federal structure, as in the US and Australia, both the federal and state governments share responsibility for the oversight of mine closure and reclamation based on harmonised standards. For instance, while OSMRE is the US federal authority established under the SMCRA, legal provisions regulating surface coal-mining operations enacted by most coal-producing states are in consonance with the SMCRA, and suitably adapted to local conditions.

In accordance with best practice, the reclamation bonds, which are submitted by coal mine owners to ensure successful completion of the process, are released subject to specific conditions. For instance, in the US, the bond release process is carried out in three phases: after establishing final topography and drainage; after ensuring re-vegetation and controlling erosion; and after meeting all regulatory requirements (US Code: Title 30: Mineral Resources, 2017a). At all times, OSMRE has access to sufficient funds to finance mine closure activities in case of default by the mine owner. In 2016, OSMRE initiated the statutory rule-making process to amend provisions applicable to self-bonding by large American coal companies, which saw a spate of mine closures driven by low coal prices (Institute for Policy Integrity, 2016).

Countries with best practices not only ensure finance for closure and restoration of mines currently in operation, but also for restoring abandoned mines. For instance, in the US, one of the key aspects of the OSMRE's mission is to mitigate the effects of past mining by aggressively pursuing reclamation of abandoned coal mines (OSMRE, 2016). A reclamation fee is levied at a rate of 28 cents per ton of coal (1 US ton equals 0.907 metric ton) produced by surface coal mining (US Code Title 30: Mineral Resources, 2017b), with lower rates for coal produced by underground mining. Another example of how finances for reclamation of abandoned mines are addressed is the Mining Rehabilitation Fund of Western Australia established in 2012, which accumulates funds collected through a levy on operating mines based on their disturbance area, for rehabilitating abandoned and future mine sites, when all other avenues to ensure their rehabilitation have been exhausted (Government of Western Australia, 2016).

Since the enactment of their SMCRA 1977, coal mine owners in the US have restored more than one million hectares of mined-out land for various beneficial uses (Deutsch & Stahl, 2017). In addition, the federal and state governments have facilitated restoration of more than 40,500 hectares of coal mines abandoned before the enactment of the SMCRA (Deutsch & Stahl, 2017).

In advanced coal-mining countries, the peculiar conditions of the mining industry have been more widely recognised, namely that surface mines require a coherent approach to regulation of mining operations, even if this involves special treatment by their respective environmental authorities. Different government agencies often work together in these countries to simplify the law. For example, the Mine Closure Guidelines in Western Australia were drafted and notified jointly by the Director General of Mines and the Chairman of the State Environmental Pollution Authority (Department of Environment and Heritage Protection, 2014). In the US, four federal agencies – US Army Corps of Engineers, US Environmental Protection Agency, US Fish and Wildlife Service and US OSMRE – signed a Memorandum of understanding in 2005 to develop collaborative processes that emphasise early and close inter-agency coordination, while maintaining their independent jurisdictional roles, as part of a jointly agreed framework to improve permit application procedures for certain surface coal-mining operations (OSMRE, 2005).

In another instance of good practice, in 2015, the Planning Commission of the Republic of South Africa declared that it was essential to simplify its policy, law and regulatory approach to the mining industry, and develop an integrated approach to mining, water and environmental statutes, to enable the country to accelerate progress toward reducing poverty, inequality and joblessness (Fischer, 2015). To address these objectives, the government created a 'One Environmental System' to develop a more coherent and predictable regulatory framework, designed to reduce red tape and the cost of compliance, while ensuring an appropriate standard of environmental performance (Fischer, 2015).

Applying good practice to India's mining closure process

According to international best practice, the assessment of social and environmental impacts, their progressive mitigation, and the review of mine closure and rehabilitation plans are carried out most effectively before and during the process of public consultation so that the community accepts the final outcomes. Currently, the approved Mine Closure Plans for coal mines in India do not incorporate any inputs from local communities or other major stakeholders since the approval of the Mine Closure Plan pre-dates the public consultation process required to obtain an Environmental Clearance as per the Environmental Impact Assessment (EIA) Notification of September 2006 (MOC, 2013) (see Hochstetler, 2017, for an example of the issues raised by the EIA in Brazil). Neither the guidelines issued by MOC nor any other statute in India stipulates stakeholder involvement during the preparation and/or approval of the Mining Plan, which now includes the Mine Closure Plan. Furthermore, no provision is made in India for the mine owner to hand over the land used for mining and allied activities to the population or to the local communities affected by the project, after reclamation and restoration of the mining-affected areas. In line with best practice evidence, the Mine Closure Plan must be given due importance in the environment clearance process to incorporate the priorities of the local communities through the process of public consultation.

India has no defined standards to assess objectively whether the mine closure has met the statutory requirements. Although the Mine Closure Guidelines mandate the handover of mined-out land to the government post mine closure and rehabilitation, no standards or procedures have been specified before the mine owners can hand over the mined-out land to the government. Since coal production is included in the Union List of the Constitution of India, the Government of India alone has the powers to regulate the coal industry. Further, three central government companies produce more than 90–95 per cent of the coal and lignite. If India is to create its own scientific knowledge and standards database for operationalising mine closure and rehabilitation for opencast mines, in line with best practice, mining companies, particularly government-controlled miners like CIL and SCCL should be required to conduct detailed scientific studies to monitor the reclamation and restoration activities in closed as well as operating coal mines.

In India, once the mining plan is approved by the coal ministry, the owner applies for environment and forest clearances to the designated authorities in the environment and forest ministry under the Environment Protection Act and Forest Conservation Act, respectively. These approvals are granted through independent processes at the state and central levels. In addition, the owner must secure various permissions from the State Pollution Control Board and other designated authorities, for example, under the Water Act, Air Act, Explosives Act and Mines Act, before applying to the Coal Controller for the final permission to open the mine.

Due to the lack of an effective consultative mechanism between the three major permitting authorities (coal ministry, environment wing, and forest and wildlife wing in the Ministry of Environment, Forest and Climate Change) within central government, opportunities to enable optimal post-mining land uses or to minimise adverse environmental impacts are missed. In some cases, to comply with the conditions imposed by different authorities, sub-optimal solutions are implemented due to the lack of an integrated approach to, resource extraction, ecological protection and community acceptance. These anomalies also arise due to the time lag between the independent review and approval processes followed by these authorities. Finally, the multiplicity of government authorities involved delays the entire process, thereby increasing risk and cost.

The CCO under MOC is mandated with the task of overseeing implementation of mine closure activities, which has proved to be impossible given the large number (500) of coal mines with approved closure plans (CCO, 2017). While the government has authorised certain government academic and research institutions to inspect coal mines and certify that the mine closure activities are progressing according to plan, discussions with key stakeholders in the mine closure process reveal that these institutions do not possess the dedicated manpower, multi-disciplinary expertise and infrastructure required to complement the Coal Controller in such a critical function and on such a large scale. Moreover, during these discussions, the Coal Controller admitted that the CCO is not equipped to reclaim and restore these mining areas in cases where the mine owner fails to discharge his responsibility as per the Mine Closure Plan. To resolve the issues of multiplicity of laws, authorities and inadequate capacities, the Government of India could consider enacting a separate and all-inclusive law for surface coal mining overseen by a unified and empowered regulator with multi-disciplinary expertise and resources, as in the case of the US.

The current statutes in India do not mandate a detailed assessment of the mine rehabilitation liability, which varies from mine to mine depending on the geo-mining conditions, mining operations, reclamation and rehabilitation activities specified in the Mine Closure Plan, as well as local conditions. Instead, the amount of INR 600,000 (approximately 6200 GBP) to be deposited by the mine owner into an Escrow account operated by the Coal Controller is applied at a flat rate (with due adjustment for inflation since 2009), irrespective of geo-mining and other conditions (MOC, 2013), which vary from mine to mine. To overcome these lacunae, it should be possible for the Coal Controller (or a future unified regulator) to estimate the cost of mine closure and reclamation on a scientific basis for each mine, in line with best practices in the US and Australia. Once such an assessment has been carried out, the Escrow amount could be set at 100 per cent of the current liability assessment to implement the Mine Closure Plan, as in the US.

Furthermore, as prescribed in the Mine Closure Guidelines in India, based on the mine owner's implementation of the Progressive Mine Closure Plan, up to 80 per cent of the total amount deposited in the Escrow account may be released every five years, without defining clearly the various milestones and standards of performance to be achieved progressively by the mine owner (MOC, 2013). Such laxity exposes the

Government of India to the uncertainties and risks of being left with a closed or abandoned coal mine without creating self-sustaining ecosystems as envisaged in the Guidelines. To avoid such a situation, the Government of India could start releasing Escrow amount in stages, only after establishing final topography and drainage in surface coal mines, again drawing on good practice in the US.

Since 2010, the Government of India has been levying a 'Clean Energy Cess' on coal and lignite used in India, whether by domestic production or through imports. It re-doubled the cess to INR 400 (4.8 GBP) per ton after renaming it 'Clean Environment Cess' with effect from 2016 (Central Electricity Regulatory Commission, 2017). The cess was subsumed into the GST compensation fund from April 2017 (MLJ, 2017). However, part of the cess could usefully be deployed for the reclamation and restoration of the 198 abandoned coal mines identified by MOC. Such a solution would fall completely within the control of the Government of India.

Conclusion

While surface mining of coal has grown markedly in India, environmental and social awareness has also been increasing. Surface coal mines have a far-reaching, impact on the environment and local communities. Over time, the capacity to rehabilitate mining-affected areas has become fundamental to the coal industry's social licence to operate. Greater public awareness has resulted in ever-increasing demands for sustainable development, which has consequently been written into the Law.

This article has analysed the process of surface coal mine closure in India in relation to best practices in some of the advanced coal-producing countries elsewhere in the world. Drawing on the evidence from other countries, a six-pronged recommendation is proposed to overcome the policy gaps identified in India, comprising:

- enactment of a separate and comprehensive law encapsulating different statutes and guidelines governing surface coal mining;
- instituting a unified and empowered regulatory body equipped with multi-disciplinary expertise, and charged with the sole responsibility for granting all clearances required for surface coal mining;
- undertaking scientific studies to create knowledge and develop standards in the Indian context for operationalising closure and restoration of mining-affected areas to create a self-sustained ecosystem;
- giving due prominence to the Mine Closure Plan in the environment clearance process involving mandatory public consultations;
- laying down rules to determine the Escrow amount for each mine on a scientific basis and linking the progressive release of this amount to the progress achieved in meeting specific requirements of the approved Mine Closure Plan and
- utilising part of the Clean Environment Cess for restoration of abandoned coal mines.

While this article was being drafted, the Government of India was reviewing the effectiveness of its policy framework for mine closure with the overarching objective of ensuring the sustainable development of the coal-bearing areas and the coal sector. In March 2017, the authors were entrusted with a research project, requiring them inter alia to engage with the government and other key stakeholders to assess the feasibility of implementing proposals such as these, and to suggest suitable technologies, operating practices, policies and/or other measures that would ensure the Indian coal-mining system is designed to be self-sustaining ecological system.

The legal mandate to ensure that mined-out land is restored to the benefit of the local communities is in the spirit of the SDGs adopted by India in 2015. A useful rule implemented in countries like the US and Australia is that, if an adequate closure plan cannot be developed, then the mine should not be opened. As a custodian of natural resources, the Government of India is obligated, henceforth, to protect fragile resources and promote community welfare by updating its mine closure policies and practices.

Acknowledgements

The authors wish to thank the reviewers and the editors of the themed issue for their insightful thoughts and comments on this article.

Disclosure statement

No potential conflict of interest was reported by the authors.

Funding

The authors acknowledge the support received from Science and Engineering Research Board, Department of Science and Technology, Government of India [SB/IR/NIAS/2016] for the first author's travel for discussions with key stakeholders involved in the mine closure process in India.

Notes on contributors

Raman Srikanth is Professor and Head of the Energy and Environment Research Program at the National Institute of Advanced Studies (NIAS), Bengaluru, India. He graduated from the Indian School of Mines in 1984, and received their Gold Medal. He worked in coal mining for six years before commencing research on projects funded by the US Bureau of Mines at Pennsylvania State University, where he obtained a dual PhD in Mining Engineering & Operations Research in 1996. He joined NIAS in 2016 after 20 years in the industry, primarily in Tata Steel. His research interests include energy studies, sustainability, minerals and management.

Hippu Salk Kristle Nathan is an Assistant Professor in the Energy and Environment Program at the NIAS, Bengaluru, India. He holds an MTech in Thermal Engineering from the Indian Institute of Technology, Delhi, and PhD in Development Studies from the Indira Gandhi Institute of Development Research, Mumbai. In 2017, he was the Subir Chowdhury Fellow on Quality and Economics at the London School of Economics. He won the Elsevier–National Academy of Sciences, India Young Scientist Award 2017 in Social Sciences. His research interests include economic measurement, energy, human development and sustainable use of resources.

References

BP. (2017). *Underpinning data 1956–2016: BP statistical review of world energy June 2017* (66th ed.). London: Author.

Central Electricity Authority. (2016). *Draft National electricity plan. Vol. 1: Generation*. New Delhi: CEA, Ministry of Power, Govt. of India. Retrieved from www.cea.nic.in/reports/committee/nep/nep_dec.pdf

Central Electricity Authority. (2017, May). *Monthly report*. CEA, Ministry of Power, Govt. of India. Retrieved from http://www.cea.nic.in/reports/monthly/executivesummary/2017/exe_summary-05.pdf

Central Electricity Regulatory Commission. (2017). Order date: 7th April 2017, Petition No. 112/MP/2015. Retrieved from http://www.cercind.gov.in/2017/orders/112.pdf

Central Pollution Control Board. (2011). *Impact of coal mine waste water discharge on surroundings with reference to heavy metals*. Bhopal: Author. Retrieved from http://cpcb.nic.in/CoalMineWW2011.pdf

Coal Controller's Organisation. (2017). *Coal Directory of India 2015–16: Coal statistics*. Kolkata: CCO, Ministry of Coal, Government of India. Retrieved from http://www.coalcontroller.gov.in/writereaddata/files/Coal%20Directory%20of%20India%202015-16.pdf

Commonwealth of Australia. (2016). *Mine rehabilitation: Leading practice sustainable development program for the mining industry*. Canberra: Author. Retrieved from https://industry.gov.au/resource/Documents/LPSDP/LPSDP-MineRehabilitationHandbook.pdf

Department of Environment and Heritage Protection. (2014). *Guideline, rehabilitation requirements for mining resource activities*. Queensland Government. Retrieved from https://www.ehp.qld.gov.au/assets/documents/regulation/rs-gl-rehabilitation-requirements-mining.pdf

Deutsch, J. R., & Stahl, P. D. (2017). *Reclaiming energy's footprint: Restoring land after coal, oil, & gas development*. Science Policy Office, Soil Science Society of America. Retrieved from https://www.soils.org/files/science-policy/caucus/briefings/reclamation.pdf

East Germany's old mines transformed into new lake district. (2016, September 7). *The Guardian*. Retrieved from https://www.theguardian.com/travel/2016/sep/17/lusatian-lake-district-project-east-germany

Fischer, D. (2015). *One environmental system*. Presentation at IAIA Annual Conference. Retrieved from http://www.elasa.co.za/uploads/1/1/8/2/11823994/d.fischer.pdf

Government of Western Australia. (2016). *Abandoned mines policy*. East Perth: Department of Mines and Petroleum, Government of Western Australia. Retrieved from http://www.dmp.wa.gov.au/Documents/Environment/ENV-MEB-201.pdf

Guha, B. (2002). The coal mining industry. In V. B. Singh (Ed.), *Economic history of India 1857–1956* (pp. 303–325). New Delhi: Allied Publisher Private Ltd.

Hochstetler, K. (2017). Environmental impact assessment: Evidence-based policymaking in Brazil and Latin America. *Contemporary Social Science*.

Institute for Policy Integrity. (2016). *Self bonding in an era of coal bankruptcy: Recommendations for reform*. New York: The Institute for Policy Integrity, New York University School of Law. Retrieved from http://policyintegrity.org/files/publications/Coal_Self-Bonding_Report.pdf

International Council on Mining and Metals. (2008). *Planning for integrated mine closure: Toolkit*. Retrieved from https://www.icmm.com/website/publications/pdfs/310.pdf

International Energy Agency. (2016). *World energy outlook 2016*. Paris: Author. Retrieved from https://www.docdroid.net/IOBt86G/world-energy-outlook-2016.pdf#page=618

Ministry of Coal. (2013, January 7). *Guideline for preparation of mine closure plan*. New Delhi: Government of India. Retrieved from http://www.cmpdi.co.in/env/MCP%207_1_2013.pdf

Ministry of Coal. (2016). Data of coal mines, Unstarred question no. 280, Lok Sabha, to be answered 25.02.2016; MOC, Govt. of India. Retrieved from http://164.100.47.190/loksabhaquestions/annex/7/AU280.pdf

Ministry of Coal. (2017). *Annual report (2016–2017)*. MOC, Govt. of India. Retrieved from http://coal.nic.in/sites/upload_files/coal/files/coalupload/AnnualReport1617.pdf

Ministry of Environment, Forest and Climate Change. (2016). Ecological impact of mining, Starred question no. 334, Lok Sabha, Dt. 09.08.2016, MOEF&CC, Govt. of India. Retrieved from http://www.indiaenvironmentportal.org.in/files/file/Ecological%20Impact%20of%20Mining.pdf

Ministry of Information and Broadcasting. (1990). *Economic structure of India, Vol. 3, The Gazetteer of India*. New Delhi: Government of India.

Ministry of Law and Justice. (2015). *The mines and minerals (development and regulation) amendment act 2015*. New Delhi: Government of India. Retrieved from http://www.indiacode.nic.in/acts-in-pdf/2015/201510.pdf

Ministry of Law and Justice. (2017). *The goods and services tax (compensation to states) act 2017*. New Delhi: Government of India. Retrieved from http://gstcouncil.gov.in/sites/default/files/GST%20(Compensation%20to%20States)%20Law.pdf

Office of Surface Mining Reclamation and Enforcement. (2005). Agencies agree to joint regulatory framework for processing applications for surface coal mining operations, News release, US Office for Surface Mining. Retrieved from https://www.osmre.gov/resources/newsroom/News/2005/021005.pdf

Office of Surface Mining Reclamation and Enforcement. (2016). OSMRE vision and mission. Retrieved from https://www.osmre.gov/about/MissionVision.shtm

Office of Surface Mining Reclamation and Enforcement. (2017). Guidelines, handbooks, manuals, and public domain software. Author. Retrieved from https://www.osmre.gov/resources/library/ghm.shtm

Ontario Mining Act. (2005). *Mine closure plans: Ontario mining act fact sheet*. The Canary Research Institute for Mining. Retrieved from http://www.canaryinstitute.ca/publications/Ontario_Closure_Brochure.pdf

Raj, B., Rethinaraj, T. S. G., & Nathan, H. S. K. (2016). Social dimension of India's energy future. *RITES Journal*, *18*(1), 12.1–12.20.

Satyanarayana, I. (2012). Basics of open pit mining, SCCL. Retrieved from https://www.slideshare.net/isnindian/basics-of-openpit-mining

Sharma, N. K. (2000). Coal mining industry in India: Past, present, and future. In A. K. Ghosh & B. B. Dhar (Eds.), *Mining challenges of the 21st century* (pp. 175–182). New Delhi: APH Publishing Corporation.

Supreme Court of India. (1996). M.C. Mehta vs Kamal Nath & Ors on 13 December, 1996, Author: Singh K. Retrieved from https://indiankanoon.org/doc/1514672/

Surface Mining Control and Reclamation Act. (1977). Public law 95-87. Author. Retrieved from https://www.osmre.gov/lrg/docs/SMCRA.pdf

United Nations Development Programme. (2015). *Sustainable development goals booklet*. New York, NY: Author. Retrieved from http://www.undp.org/content/dam/undp/library/corporate/brochure/SDGs_Booklet_Web_En.pdf

US Code: Title 30: Mineral Resources. (2017a). *Performance bond, financial assurance, and insurance requirements for surface coal mining and reclamation operations*. Retrieved from https://www.ecfr.gov/cgi-bin/retrieveECFR?gp=&SID=0681606cf267ddbe91be104a2601d27d&mc=true&n=pt30.3.800&r=PART&ty=HTML

US Code Title 30: Mineral Resources. (2017b). *Abandoned mine reclamation fund – fee collection and coal production reporting*. Retrieved from https://www.ecfr.gov/cgi-bin/retrieveECFR?gp=&SID=f47c1f42bc955be994684dc5c45b6ae8&mc=true&n=pt30.3.870&r=PART&ty=HTML

World Commission on Environment and Development. (1987). *Report of the world commission on environment and development: Our common future*. London: WCED, UN. Retrieved from http://www.un-documents.net/our-common-future.pdf

World Development Indicators. (2017). *Time series data for countries* (last updated April 27, 2017). Retrieved from http://data.worldbank.org/data-catalog/world-development-indicators

Zhao, L. Y. L., McCullough, C. D., & Lund, M. A. (1999). *Mine voids management strategy (I): Pit Lake resources of the Collie Basin*. Joondalup, WA: Center for Ecosystem Management. Retrieved from https://www.water.wa.gov.au/__data/assets/pdf_file/0012/5601/96275.pdf

The impact of urbanisation on land resources in Chongqing, China, 1997–2015

Huiming Zong and Bingjie Cai

ABSTRACT

Significant differences are observed in the urbanisation process between China and countries in the Western world, and also within China between its western regions and coastal areas. The western region of China experienced rapid urbanisation after 1997 when Chongqing, one of the most important central cities in the region, came directly under central government control. This article examines the urbanisation process in Chongqing from 1997 to 2015. The authors use the coordinated degree index, land use change intensity index and land use spatial index to analyse the urban construction process, its spatial variability and impact on cultivated land resources from a human geographer's perspective. They show how rapid urban growth kept pace with the rate of industrialisation and was characterised by a 'centre–periphery' spatial structure. They find that urban growth lagged behind the rate of urban construction, measured by the rate of population urbanisation. In line with the government's 'new urbanisation strategy', which was supported by research evidence, arable land was not used for construction, in contrast to other coastal cities. In identifying the characteristics of the urbanisation process in Chongqing and the challenges it raised, the authors comment on the reciprocal relationship between research evidence and policy development.

Introduction

Urbanisation is a dynamic process, encompassing urban and rural population and employment transfers, lifestyle changes and the evolution of urban–rural spatial structures. Currently, China is undergoing a critical period of urban transformation. The process is being driven by an economic structure dominated by traditional agriculture with priority being given to modern industrial transformation (Wang, Liu, & Li, 2015). The measurement and process of urbanisation at national level and in coastal areas has been widely discussed by urban geographers, who highlight the characteristics of the rapid growth of the urban population and the extension of the urban area (Deng, Wang, Hong, & Qi, 2009; Lin & Ho, 2003).

With only 7% of the world's arable land and 22% of the world's population, it is especially important for China to protect its arable land (Huang, Zhu, Qiao, & Sun,

1998), the more so since the loss of some non-agricultural and cultivated land resources during urbanisation is irreversible (Liu, Xu, Liu, Chen, & Ning, 2015). Some measures have been taken to increase economic revenue and enhance sustainable development at the local level by reforming land management systems (Li & Zhou, 2005; Tao, Su, Liu, & Gao, 2010; Wu, Li, & Yan, 2015) to make urban space construction less intensive, to prevent uncontrolled urban sprawl and to overcome obstacles to the land transfers process (Lu & Ye 2007). Drawing on research evidence, the Government of China has recognised that it is essential for the country's economic and social sustainable development to achieve the targets set for the protection of land resources, especially cultivated land, during the urbanisation process, as set out in the United Nation's Sustainable Development Goals (2, 11 and 15).

With a view to informing policy and ensuring sustainability of resources, the academic community in China has conducted extensive research into the impact of urbanisation on cultivated land. Their work has shown, firstly, how the development of urbanisation has encroached on cultivated land, resulting in a decline in both the amount of cultivated land (Feng & Li, 2000) and in its quality (Liu, Liu, & Tian, 2005). Secondly, using empirical studies, some researchers have found that higher levels of urbanisation can ease the pressure on cultivated land and help protect it. Such evidence is influential since it demonstrates how governments in other urbanised areas have recognised the need to protect cultivated land (Liu et al., 2015). Thirdly, it has been argued that the different stages of urbanisation may have different impacts on cultivated land: land use in small towns is less intensive, whereas the occupation of cultivated land in larger agglomerations may be too dense; and intervention at the right stage of development of big cities may help to protect arable land (Cai & Zhou, 2005). A fourth perspective suggests that the extension of big cities is the main cause of the reduction in cultivated land (Pan & Liu, 2014). China covers such a vast territory and such a large number of diverse administrative units that local urbanisation practices are rarely identical (Bai, Shi, & Liu, 2014). In other words, no consensus exists regarding the possible impact of urbanisation on cultivated land resources and the ways in which governments can intervene to protect natural resources.

In the context of the Chinese Government's 'new urbanisation strategy' and its impacts on land resources, the municipality of Chongqing was selected for a case study, due to the urgent need to understand the characteristics of the urbanisation process in a mountainous area in western China. Chongqing has been the economic centre of the western region of China and the upper reaches of the Yangtze River since the 1970s. It is also an important land and water transportation hub and foreign trade port. From being under the jurisdiction of Sichuan province, it was brought directly under the jurisdiction of the central government in June 1997. The urbanisation process in Chongqing took place in the context of globalisation, industrialisation, and new urban strategies from both central and local governments, which distinguishes it from the situation of Western countries and the eastern coastal region in China.

Based on the development and characteristics of urbanisation in Chongqing, this article analyses the impact of urbanisation on land use from the perspective of both urban construction and cultivated land, with the aim of providing evidence about the most cost-effective ways of controlling the development of urbanisation and managing land resources, for subsequent application in other regions of China facing similar problems.

The evidence base for tracking urbanisation

The data used in this study come from the *China statistical yearbook*, *China population employment statistics yearbook*, *Chongqing statistical yearbook*, *Chongqing survey yearbook*, and *The land change survey*. Data from the official yearbooks that are published by government departments have official authority. These data have been used to track changes in population, economy and development at national and regional levels. ARCGIS software was used for data processing and image visualisation, covering Chongqing and its 38 districts.

The quantitative parameters selected to study land use change include: the land change intensity index (Elli), and the land use spatial differentiation index (G) (Ji, Yang, & He, 2014). The deviation between population urbanisation and land urbanisation (C_v) is used to measure the imbalance between population urbanisation and the expansion of urban construction land (Pan & Liu, 2014). The coordination degree index (C_v) is a measure of coordination between two systems or elements. It reflects the relative degree of variation between the two sets of data and is especially applicable to the degree of variation between the different data in the measurement unit (see Table 1).

Elli refers to the standardisation process for the annual average rate of expansion by the spatial unit land area. This indicator renders the urban land expansion rate comparable in different periods. If the value of G is positive, the indicator shows that the county is following a similar trend to the city; if the value of G is negative, the region is experiencing the opposite overall trend, and the region is contributing less to the trend in regional change. The larger the C_v, the greater the deviation between population urbanisation and the expansion rate of urban construction land, and the larger the mismatch between population urbanisation and the expansion of urban construction land. The smaller the C_v, the more harmonious the development of population urbanisation and the expansion of construction land.

Table 1. Parameters, meanings and formulae for urbanisation indices.

Parameter	Meaning	Formula
Land change intensity index	To reflect the change intensity of certain types of land at different periods	$\text{Elli} = \dfrac{A_t - A_0}{T \times s} \times 100\%$
Land use spatial differentiation index	To reflect the change in land use for spatial units, partly reflecting the contribution of a region to the overall trend for the type of change	$G = \dfrac{A_t - A_0}{U_t - U_0} \times 100\%$
Coordination degree index	To reflect the pace of urbanisation and the extent of the expansion of urban construction land	$C_v = \dfrac{\sigma}{\lvert\bar{X}\rvert} = \dfrac{\sqrt{\frac{1}{2}\left[\left(P - \frac{P+L}{2}\right)^2 + \left(L - \frac{P+L}{2}\right)^2\right]}}{\left\lvert\frac{P+L}{2}\right\rvert} = \left\lvert\dfrac{P-L}{P+L}\right\rvert$

Notes: A_0 and A_t are the areas of a certain type of land at the beginning and end of the study, T is the research period of time, S is the total area of land in the study area, G is the land use spatial differentiation index, U_0 and U_t are the whole area of a certain type of land at the beginning and end of the study. C_v is the dispersion coefficient, σ is the standard deviation, $\bar{X}$ is the average; P stands for the growth rate of population urbanisation and L for the growth rate of urban construction land.

Characteristics of the urbanisation process in Chongqing

Our analysis shows that Chongqing experienced an above-average urbanisation growth rate compared to the national average during 1997–2016. The high growth rate in urban development was driven by large infrastructure investment and the good development prospects of dominant industries, including the automobile, electronic and communication industries. Chongqing entered a rapid urbanisation stage with a rate of over 30% in 1997. At that time, the urban population of Chongqing stood at 8.91 million, and the population urbanisation rate was 31% and had risen to 62.6% in 2016, with an average increase of 1.7 percentage points per year. Comparatively, the urbanisation rate nationwide was almost the same as in Chongqing in 1997, but the national rate increased by only 1.3 percentage points per year between 1997 and 2016, as shown in Figure 1.

Figure 2 demonstrates how the trend in population outflow from Chongqing was gradually slowing down, while the inflow of population from outside was increasing. Since the opening up of the Chinese economy under the 1978 'Reform of the economic system', as in the most western part of China, Chongqing became one of the main regions from which the population was migrating to large cities in the coastal area. Until 2004, the total number of permanent residents in Chongqing was declining by about 110,000 inhabitants per year, as the result of population migration to coastal areas from western parts of China. From 2004, the resident population in Chongqing increased gradually, at an average annual rate of around 200,000 people. This phenomenon is explained by the growing attractiveness and the expanding agglomeration capacity of Chongqing city, whereas the coastal cities were facing economic problems during the global financial crisis. The urban population thus grew steadily as the region became more urbanised.

Household population registration refers to citizens who, according to the 'The household registration regulations of the People's Republic of China', are registered as permanent residents at the household's usual place of residence. The permanent population refers to the population that has been living in the same place for a number of years or for more than six months in the year at the time of population census. The urban household population registered in Chongqing was relatively small, as shown in Figure 3.

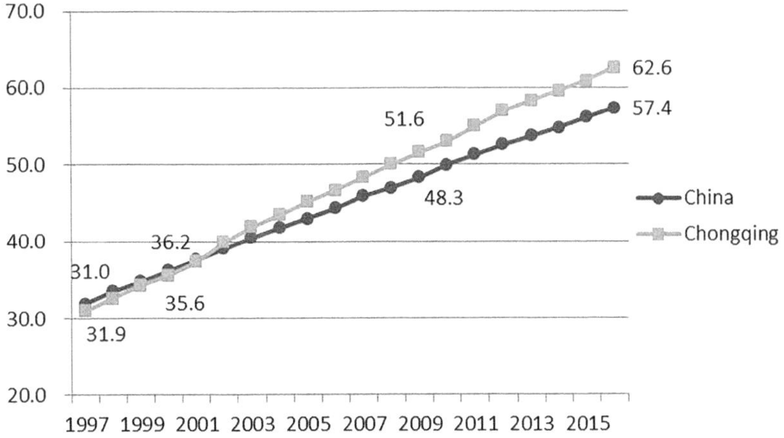

Figure 1. Urbanisation rates in Chongqing and China in 1997–2015. Source: Authors' data.

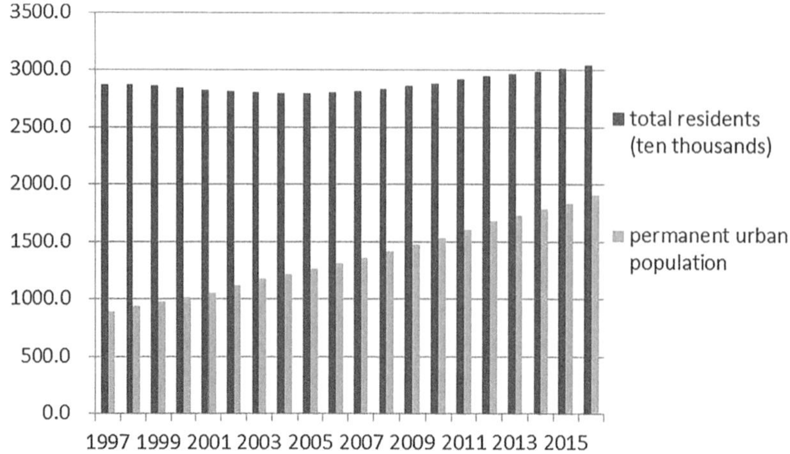

Figure 2. Total population and urban population of Chongqing in 1997–2015. Source: Authors' data.

The gap between urban and total registered population in the province narrowed briefly between 2009 and 2013, although the total number was still increasing, as shown in Table 2. The gradual relaxation of the national household registration system, combined with policies to promote urbanisation in Chongqing, increased the level of urbanisation, as more migrant workers were attracted from rural areas. Although more of the migrant population live in the urban area, most of those with rural *hukou* registration refuse to take urban *hukou* registration, because of their attachment to their home in a rural area, and to keep any potential benefits of access to land such as central government subsidies for peasant farmers. The rapid increase in the migrant population with rural *hukou* registration therefore explains the increasing gap between the number of permanent population and household population registrations.

In general, the development of urbanisation in Chongqing has been increasingly in line with trends in industrialisation, as illustrated by Figure 4. According to the international standard value method, IU is the ratio of the labour industrialisation rate to the

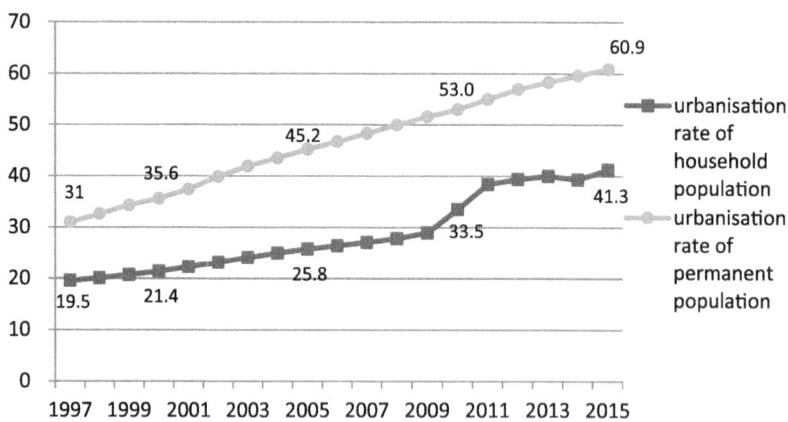

Figure 3. Chongqing household and permanent population urbanisation rates in 1997–2015. Source: Authors' data.

Table 2. Fundamental data for urbanisation and land use in Chongqing.

Year	Permanent resident population (millions)	Registered population (millions)	Urban population (millions)	Urbanisation rate (%)	Arable land (hectares)	Urban construction land (hectares)
1997	28.74	30.42	8.91	31.0	2,541,301	32,087
2000	28.48	30.91	10.14	35.6	2,522,925	43,433
2005	27.98	31.69	12.66	45.2	2,262,697	62,906
2010	28.85	33.03	15.30	53.0	2,442,867	138,416
2015	30.17	33.72	18.38	60.9	2,430,468	190,323

Source: Authors' data.

urbanisation rate. NU is the ratio of the labour non-agricultural rate to the urbanisation rate. If the IU ratio is about 0.5, and NU is about 1.2, urbanisation, industrialisation and non-agricultural development are more harmonious; with IU > 0.5 and UN > 1.2, urbanisation development is considered insufficient; with IU < 0.5, NU < 1.2, urbanisation is more advanced. In 1997, the value of the IU in Chongqing was 0.59, and NU was 1.74, indicating that urbanisation was lagging behind the development of industrialisation and non-agricultural land use (Zhu, Dai, & Xie, 2012). A large number of workers in industrial and other non-agricultural industrial production were stranded in rural areas and unable to migrate to urban areas. The IU and NU values then fell over time, as the trend in urbanisation development ceased to lag behind the industrialisation process and moved closer to that for coordinated industrialisation and non-agricultural development.

The spatial configuration of population urbanisation

The 38 counties in Chongqing can be divided into four categories based on urbanisation levels: lower (below 30%), low (30–50%), high (50–70%), higher (above 70%). Analysis of spatial differences in urbanisation level in each county and district of Chongqing showed that the level was decreasing from the central city to the northeast and southeast of Chongqing, revealing a 'centre–periphery' structure, as illustrated in Figures 5 –7. The districts around the urban centre displayed a higher level of urbanisation in 2015; the area surrounding the city centre was in the high range. The western part of Chongqing was mainly the ecological conservation development and protection zone in northeast

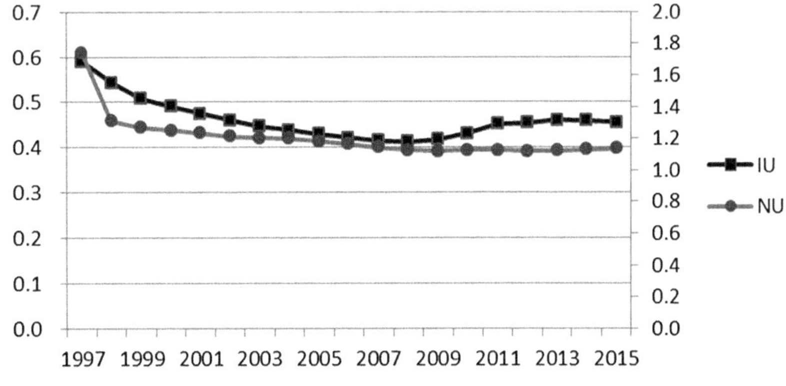

Figure 4. IU and NU value of Chongqing in 1997–2015. Source: Authors' data.

Chongqing, including parts of the Three Gorges Reservoir area, where it displayed a level in the lower range. The geographical distribution of the city can be seen to be gradually shifting from the city centre to the surrounding area.

Spatial expansion thus follows a general trend of central radiation and gradient progression. The level of urbanisation in the city centre and its surrounding counties was higher, thereby stimulating the development of both surrounding counties and districts. Hechuan, Yongchuan and Bishan Districts benefitted from radiation from the urban centre and were in the medium level of development, with an urbanisation rate roughly equivalent to the average level for Chongqing. Besides Wanzhou District (the leading district in northeast Chongqing), southeast Chongqing and northeast Chongqing, which were farther away from the metropolitan area, generally experienced a below average level of development. Thus, the pattern of radiation gradually decreased from the urban centre to surrounding areas.

Urbanisation and urban construction land

As Chongqing entered a period of rapid development from being a provincial municipality, considerable pressure was exerted on land resources (Braimoh & Onishi, 2007). As noted above, the rate of urbanisation in Chongqing lagged behind the expansion rate of urban construction land. The dispersion coefficient of the urban population and urban construction land in 1997–2015 was 0.65, which can be considered as highly disordered (see Table 3). At the beginning of 1997, the total number of permanent urban

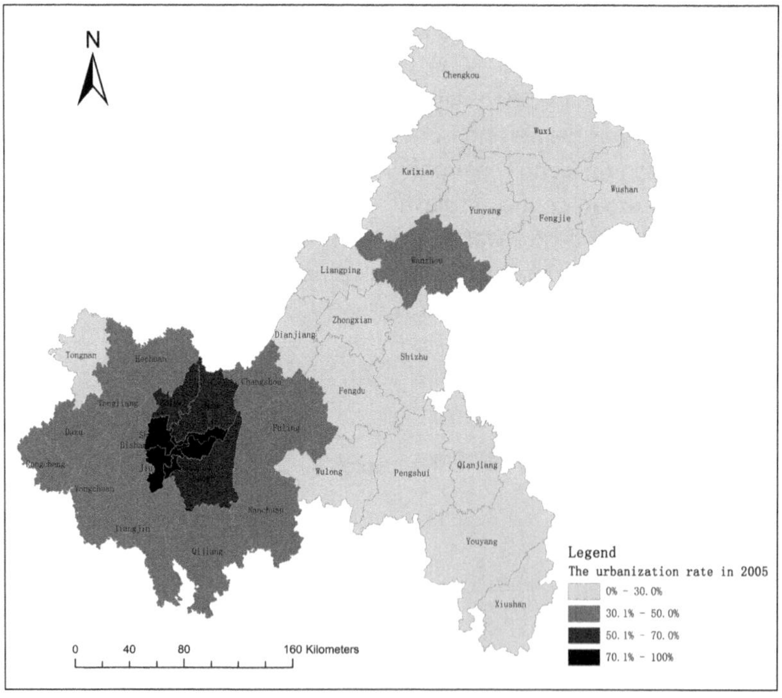

Figure 5. Spatial distribution of urbanisation in Chongqing municipality in 2005. Source: Authors' data.

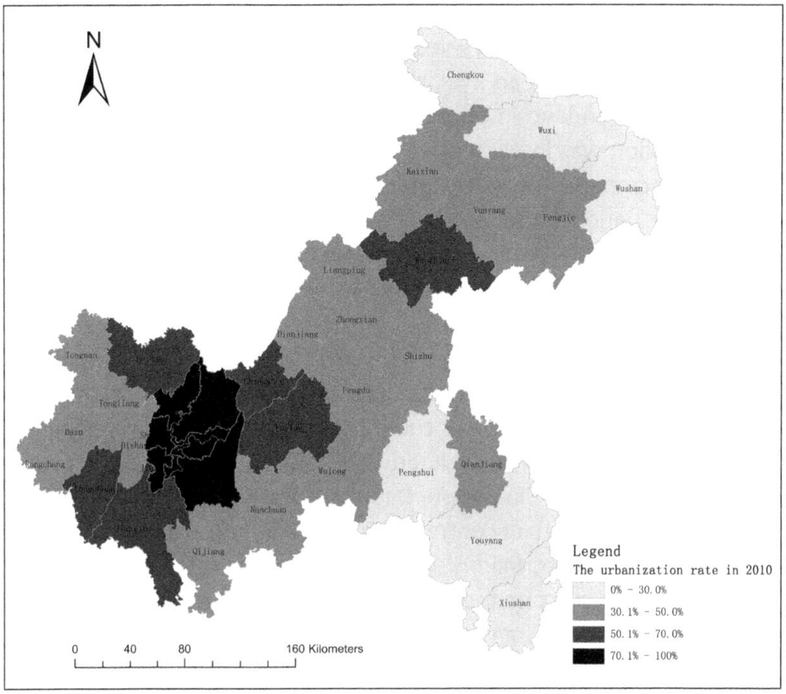

Figure 6. Spatial distribution of urbanisation in Chongqing municipality in 2010. Source: Authors' data.

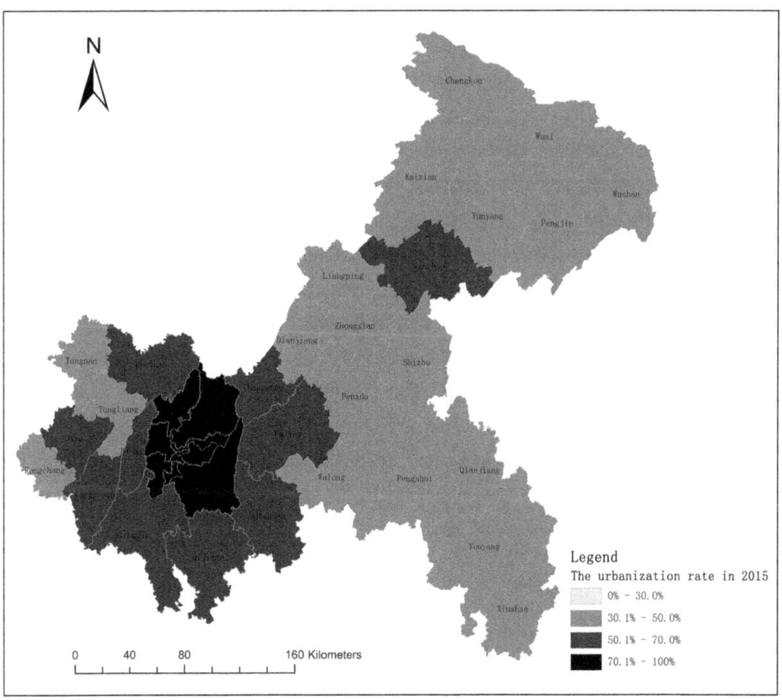

Figure 7. Spatial distribution of urbanisation in Chongqing municipality in 2015. Source: Authors' data.

residents was 8,907,400 and, by the end of 2015, it had risen to 18,384,100. The urban population increased 1.06 times in 19 years. During the same period, urban construction land expanded 4.93 times, with an average annual increase of 8328 ha. Since the expansion of urban construction land in Chongqing was faster than urban population growth, urban land construction was less intensive during the process of rapid urbanisation.

The degree of discrepancy between urbanisation and construction land in Chongqing declined during the period under study (1997–2015), and the maladjustment situation stabilised. The coordination proportion of population and construction land between 2011 and 2015 was between 0.2 and 0.4, which corresponded to a mild-disorder state, thus indicating an improvement.

Most counties in Chongqing displayed a disorder level between urbanisation and construction land, mainly due to the lag in population urbanisation. In Chongqing municipality, 8 districts and counties were at the coordinated level, and the remaining 30 districts were all disordered. The highly disordered level districts were mainly concentrated in the urban centre, such as the Jiulongpo, Dadukou and Shapingba Districts. The moderate-disorder area was in the urban centre and its surrounding area, including the Jiangbei, Banan and Nanchuan Districts. Mild-disorder type districts were mainly concentrated in the southeast and northeast Chongqing, including Wanzhou, and Fengjie Districts, and Youyang County. The metropolitan areas were the main areas for the economic development of Chongqing, and the urban areas attracted industrial relocation. Industrial, commercial, manufacturing and service industries needed large quantities of urban construction land, with the result that the rate of expansion of urban construction land was greater than that of population urbanisation.

The construction rate of cities in the ecological and protection zones was relatively slow, and the degree of disorder between urbanisation and urban construction land was low. The degree of disorder in these areas and zones was, therefore, higher than it would otherwise have been, especially for moderate- and high-disorder districts. However, with the promotion and more intensive reform of the land and household registration systems, as presented below, the disorder level of each county in Chongqing decreased, and the maladjustment situation was relieved.

Four counties belonging to the category of coordinated development accounted for 10% of the spatial distribution, while the highly, severely and extremely disordered counties accounted for 47.3%. By 2015, the number of counties with coordinated development increased to 10, accounting for 26.3% of the total, and no seriously disordered imbalanced counties were recorded. The number of counties categorised as highly and extremely disordered fell to 3, accounting for 7.9% of the spatial distribution.

Under the influence of central government policy, the reform of the household registration system published in 2010 lowered the threshold for the settlement of the rural transfer population and ensured the provision of basic public services. The urban agglomeration's ability to absorb the rural population improved. Construction land gradually

Table 3. Urban construction land disorder categories.

Disorder level	Coordinated	Mild	Moderate	High	Serious	Extreme
	0–0.2	0.2–0.4	0.4–0.6	0.6–0.8	0.8–1	>1

Source: Yin and Xu (2013).

shifted from simple incremental expansion to inventory optimisation, and the urban sprawl was restrained to a certain extent. As a result of policy changes, the disorder ratings for the whole county greatly improved, and the levels of disorder were reduced to mild and moderate.

Urbanisation and changes to cultivated land

The American geographer Northam (1979) divided the process of urbanisation of Western countries into three stages: the initial stage of urbanisation, where the level of urbanisation is less than 30%; the acceleration stage of urbanisation with between 30% and 70%; and the stable stage of urbanisation with over 70%. Northam suggested that the three stages would present a process of slow to fast to slow urbanisation, expressed as an *S* curve.

In the case of Chongqing City, as urbanisation has intensified, the amount of cultivated land has declined. Between 1997 and 2015, the level of urbanisation increased by 29%, and the total area of cultivated land decreased by 110,833 ha. The largest fall in the amount of cultivated land area occurred between 2000 and 2005, decreasing by 260,227 ha. At this stage, the pace of urbanisation development reached its peak, increasing by 1.92% per year, while the cultivated land area decreased by 27,107 ha as Chongqing entered Northam's rapid growth stage, putting pressure on cultivated land resources, with the result that the area of cultivated land declined year on year. The rapid decrease of cultivated land in Chongqing before 2005 is found in other coastal cities, but in Chongqing the decline slowed down after 2005 under the new urbanisation strategy and polices from both central and local government.

As argued by Northam, the reduction in cultivated land due to urbanisation is a periodic phenomenon. In the Chongqing case, with the overall decline in the quantity of cultivated land from 2005 to 2010, the land change intensity index remained positive, implying that the quantity of cultivated land was on the increase. In addition, because the quantity of cultivated land had fallen sharply in the previous stage, and due to the lower base data, the dynamic degree of land use change at this stage was 1.59%.

While maintaining steady economic growth, Chongqing preserved and increased the existing stock of cultivated land by digging into the reserve of cultivated land. Our findings confirmed that the reduction in the quantity of cultivated land due to urbanisation should be seen as a staged phenomenon that might be expected to change in response to the reform and adjustment of agricultural and urbanisation policy.

In 1997, the land structure of Chongqing was dominated by inferior cultivated land, accounting for 69% of the surface area, mainly in the dry land area and part of Wang Tian Tian. The development of agricultural modernisation in the process of urbanisation, the progress made in land remediation, and the investment in agricultural infrastructure led to the disappearance of Wang Tian Tian, since it had no irrigated water conservancy facilities. The proportion of paddy fields and irrigated fields in the better-cultivated land gradually increased from 29.51% in 1997 to 39.85% in 2015. As the non-agricultural population increased, and the surplus rural labour population moved to the suburbs, the vegetable growing land in the suburbs gradually disappeared. Thus, when Chongqing became a provincial municipality, the total cultivated land change intensity fell by −4.36% even though the total amount of cultivated land declined.

In sum, the total amount of cultivated land declined in the process of rapid urbanisation at the provincial municipality level. The same result was found in previous urbanisation literature (Deng et al., 2009). However, two differences in the findings should be noted. Firstly, with urbanisation, the structure of cultivated land changed as the proportion of better quality cultivated land increased slightly. About 40% of the arable land in Chongqing was of relatively good quality, consisting of paddy fields, irrigated land and vegetable fields. Secondly, the trend for arable land at the county level is different from that at the municipal level. The spatial configuration of arable land at country level demonstrates that the loss of arable land in the process of rapid urbanisation can be avoided.

Strategies for sustainable urbanisation

Prior to 2005, the development of the cultivated land area in each district followed a unified trend, and the differences between counties were not obvious. Between 1997 and 2005, Chongqing underwent a period of adjustment and transition under the control of central government. This proved to be a time for deepening the understanding of the process, identifying problems and developing new policies for improving urban development in the region. After several years spent exploring the core economic regions with more marked cluster and radiating functions, regional characteristics were documented, and the location of each district and county within the process of urbanisation gradually became clearer, as did the impact of the loss of arable land caused by urbanisation.

Between 2005 and 2010, after the overall growth in the amount of cultivated land, the land area cultivated in the districts of the main city and the surrounding counties continued to decrease. In 2006, Chongqing proposed a strategy for sustainable urbanisation, described as 'One ring and two wings'. The new urbanisation policy followed the three principles: fair sharing, intensive efficiency and sustainability. To develop regional growth poles and strengthen the radiating effect of large cities on the surrounding area, the 'One ring and two wings' strategy focused on building the metropolitan area as the core of a 'One-hour economic circle'. The strategy was designed to accelerate the development of industrial bases and clusters, particularly for modern manufacturing, services and agriculture, while making the city centre more influential in surrounding regions. Thus, from 2005 to 2010, while the general trend was to promote the overall growth of the cultivated land area, the unintended consequence of urban construction was that the quantity of arable land in the 'One ring' continued to decline.

In 2013, a further strategy was proposed for the development of 'Five functional areas' in Chongqing municipality, namely, the urban function core, the urban function development area, the city development zone, Chongqing northeast ecological conservation development area and the Chongqing northeast ecological protection zones. To ensure national food security, the Ministry of Land and Resources formulated policies to 'safeguard economic growth and the red line of arable land' (see Xinhua News Agency, 2014). These policies required strict protection of farmland both at national and provincial levels to prevent the decline in the amount of arable land in the provinces.

Between 2010 and 2015, according to the 'Five functional regions' strategy, the urban centre and its surrounding districts were the main districts selected for reducing the total area of cultivated land in Chongqing. After 2010, the urban outer space of Chongqing city expanded into the outer highway ring. The core area of the urban centre began to adjust to the industrial structure. The surrounding districts and counties were the hot spots for industrial relocation from the main urban areas, such as Banan and Nanan Districts. Therefore, compared with the previous stage, the counties with less arable land were still concentrated in the main urban centre and surrounding areas, and their outreach was expanded.

The policies developed at the local level were designed to maintain a balance between urbanisation and land resources in Chongqing province. The national 'Five functional areas' strategy provided a framework for the spatial configuration of urbanisation and land use in Chongqing city, with the emphasis on the quality of urbanisation. Drawing on this strategy and on other national policies, Chongqing proposed a series of urbanisation policies to guide local practices, and these policies worked successfully together in balancing urbanisation and land resources under the cooperation of governments of city, county and village level.

Conclusion

This article has analysed the urbanisation process of Chongqing city and its impact on land resources from two perspectives: urban construction land and cultivated land. From being a provincial municipality, the urbanisation level of Chongqing entered a rapid growth stage, and urbanisation was coordinated with industrial development. The spatial characteristics of urbanisation in Chongqing illustrate the mode of central radiation and gradient, and the core–periphery structure in metropolitan and rural areas.

The rate of population urbanisation in Chongqing lagged behind the urban construction land expansion rate, resulting in a large imbalance between the two. The radiating effect of the urban construction land was much greater than the crowding effect of the population and urban sprawl, with the result that land use was affected.

The Chinese government is constantly seeking to solve the problems of urban development through collaboration between government departments and researchers. The Government requires researchers to provide reliable scientific evidence to inform government policies, designed to promote sustainable economic development. As a result of the new policies about land use, urbanisation, spatial planning introduced during the period under study, the degree of disorder of each county in Chongqing decreased, the maladjustment situation was alleviated, and the disorder between urbanisation and land use in the whole region was improved.

With the development of urbanisation in Chongqing, as the amount of cultivated land declined, its structure changed. The proportion of prime cultivated land increased slightly. During the rapid urbanisation phase, development policies regarding functional areas had differential impacts on cultivated land resources. The cultivated land area of districts and counties that gave priority to ecological concerns increased, whereas the cultivated land area of districts and counties that gave priority to industrial development declined. Thus, the reduction in the amount of cultivated land due to urbanisation was shown to be a periodic phenomenon.

The reform and adjustment of development and land use policy were found to have had marked effects on the change of cultivated land. The decrease in the proportion of cultivated land area was mainly concentrated in the metropolitan area and surrounding counties, with the result that the cultivated land area in northeast Chongqing increased. Urban expansion in the metropolitan area provided an important market for the transfer of rural to cultivated land. At the same time, policies designed to reduce population urbanisation and to curtail land loss limited the phenomenon of disorderly urban sprawl, providing further evidence of the positive guiding role of policy in the sustainable development of urbanisation and land use. The new urbanisation strategies introduced by the central government in China in conjunction with supporting policies from local government have resulted in a rebalancing of urbanisation and land resources at city and regional levels. The findings from the Chongqing case therefore afford a model that can be used by other cities in the western part of China that are experiencing a high rate of urbanisation and are seeking to promote sustainable urban development.

Acknowledgements

The author would like to thank the Guest Editors and the anonymous reviewers for their scholarly guidance and constructive comments.

Disclosure statement

No potential conflict of interest was reported by the authors.

Funding

This work was supported by National Natural Science Foundation of China [41671159].

Notes on contributors

Huiming Zong is a Professor of Human Geography in the School of Geographical Sciences, Southwest University, Chongqing, China, and a member of the Geographical Society of China. He holds a PhD from Sun Yat-Sen University, Guangzhou, China. He was a Visiting Fellow at the LSE Asia Research Centre in 2013–2014. His research interests focus on urbanisation in China, urban spatial structure and urban planning, and the role of government in the urbanisation process of Chongqing City.

Bingjie Cai is a graduate student in the School of Geographical Sciences, Southwest University, Chongqing, China, majoring in human geography. Her research interests focus on the urbanisation process and its impact in China.

References

Bai, X., Shi, P., & Liu, Y. (2014). Realizing China's urban dream. *Nature*, *509*, 158–160. doi:10.1038/509158a

Braimoh, A. K., & Onishi, K. (2007). Spatial determinants of urban land use change in Lagos, Nigeria. *Land Use Policy*, *24*(2), 502–515. doi:10.1016/j.landusepol.2006.09.001

Cai, J., & Zhou, B. (2005). On urbanization and farmland protection. *Social Sciences Shanghai China*, *6*, 5–12 [in Chinese]. doi:10.3969/j.issn.0257-5833.2005.06.001

Deng, J. S., Wang, K., Hong, Y., & Qi, J. G. (2009). Spatio-temporal dynamics and evolution of land use change and landscape pattern in response to rapid urbanization. *Landscape and Urban Planning*, *92*(3–4), 187–198. doi:10.1016/j.landurbplan.2009.05.001

Feng, Z., & Li, X. (2000). The stratagem of cultivated land and food supplies security: Storing food in land – raising the comprehensive productivity of land resource of China. *Geography and Territorial Research*, *163*, 1–5 [in Chinese]. doi:1001-8107(2000 03-0001-05

Huang, N., Zhu, Z., Qiao, Y., & Sun, D. (1998). An analysis of the changes of cultivated land in urbanization process in the pearl river delta. *Tropical Geography*, *18*(04), 296–301 [in Chinese]. doi:10.13284/j.cnki.rddl.000309

Ji, T., Yang, H., & He, T. (2014). Analysis of spatial-temporal characteristics and driving forces of urban construction land expansion in the metropolitan area of Chongqing. *Resources and Environment in the Yangtze Basin*, *1*, 60–66. doi:10.11870/cjlyzyyhj201401009.

Li, H., & Zhou, L. (2005). Political turnover and economic performance: The incentive role of personnel control in China. *Journal of Public Economics*, *89*(9–10), 1743–1762 [in Chinese]. doi:10.1016/j.jpubeco.2004.06.009

Lin, G., & Ho, S. (2003). China's land resources and land-use change: Insights from the 1996 land survey. *Land Use Policy*, *20*(2), 87–107. doi:10.1016/S0264-8377(03)00007-3

Liu, J., Liu, M., & Tian, H. (2005). Spatial and temporal patterns of China's cropland during 1990–2000: An analysis based on Landsat TM data. *Remote Patterns of Environment*, *98*(4), 442–456. doi:10.1016/j.rse.2005.08.012

Liu, L., Xu, X., Liu, J., Chen, X., & Ning, J. (2015). Impact of farmland changes on production potential in China during 1990–2010. *Journal of Geographical Sciences*, *25*(01), 19–34. doi:10.1007/s11442-015-1150-6

Lu, D., & Ye, D. (2007). *Advice on curbing aggressive urbanisation and out-of-control urban space.* Advisory Report by Academician of the Chinese Academy of Sciences [in Chinese].

Northam, R. (1979). *Urban geography* (2nd ed.). New York, NY: John Wiley & Sons.

Pan, A., & Liu, Y. (2014). The degree of imbalance between population urbanization and land urbanization of Xiangjiang River Basin. *Economic Geography*, *34*(05), 63–68 [in Chinese]. doi:10.15957/j.cnki.jjdl.2014.05.012

Tao, R., Su, F., Liu, M., & Gao, G. (2010). Land leasing and local public finance in China's regional development: Evidence from prefecture-level cities. *Urban Studies*, *47*(10), 2217–2236. doi:10.1177/0042098009357961

Wang, Y., Liu, Y., & Li, Y. (2015). Spatial–temporal patterns of urbanisation and rural development and their coordination in Bohai Rim region. *Geographical Research*, *34*(1), 122–130 [in Chinese]. doi:10.11821/dlyj201501011 1

Wu, Q., Li, Y., & Yan, S. (2015). The incentives of China's urban land finance. *Land use Policy*, *42*, 432–442. doi:10.1016/j.landusepol.2014.08.015

Xinhua News Agency. (2014). Ministry of Land and Resources, National Land General Inspector's Office of Red Line : No new construction land in megacities. *Shanghai City Planning*, *1*, 128. [in Chinese].

Yin, H., & Xu, T. (2013). The mismatch between population urbanization and land urbanization in China. *Urban Planning Forum*, *39*(02), 10–15. doi:1000-3363(2013)02-0010-06

Zhu, Y., Dai, H., & Xie, F. (2012). Evaluation and analysis of the coupling between urbanization and industrialization of Jinan city. *Areal Research and Development*, *31*(01), 70–73 [In Chinese]. doi:1003-2363(2012)01-0070-04

Negotiating transparency: NGOs and contentious politics of the Extractive Industries Transparency Initiative in Ghana

Nelson Oppong

ABSTRACT
Transparency has been upheld by the dominant *Zeitgeist* of the twenty-first century as an all-purpose recipe for addressing the ills associated with resource-led developmental transformation. However, little attention has been paid to the bargains and contestations accompanying its institutionalisation in resource-rich countries. To gain a fuller understanding of how transparency interventions interact with the deeper vectors of power and politics embedded in resource governance, this article examines the dynamics of NGO participation in the Extractive Industries Transparency Initiative (EITI), an international auditing and multi-stakeholder oversight initiative, adopted by Ghana in 2003. By recasting the analytic problem around the optic of NGO contestation and representation, the article offers a more nuanced engagement with the material politics of negotiability, contestation, and representation that drive EITI compliance, non-compliance, and de-compliance. Drawing from an overall comparative political economy approach and a heuristic model highlighting the intricate dynamics of transparency, the overriding argument ranks the understated contestations within the NGO community around the EITI and the disparity between its platforms and the representational processes of democratic accountability among the most formidable threats to the EITI's noble ambition of securing optimum resource-led transformation.

Introduction

After a prolonged period of cynicism, a newfound optimism about the prospect of commodity-led transformation has set the development community alight. This optimism hinges on a widely held presumption suggesting that the most pronounced forms of corruption and institutional fragility, often associated with resource-rich countries, do not emanate from the presence of natural resources per se, but how they are managed and used (O'Higgins, 2006). Consequently, many activists and researchers have rallied around institutional fixes that have sought to promote norms of appropriate behaviour, including public sector reforms, transparent and inclusive processes, as well as citizens' voice and public accountability, as the formula for securing optimum benefits from the extractive sector. However, the growing interest in tracking how various actors comply with these presumably settled norms and acceptable initiatives has not been accompanied by an equal measure of enthusiasm about how they interact with the deeper vectors of power, politics, and contestations. Also, beyond ground-up accounts of the community dynamics of resource governance (Agrawal & Gibson, 1999; Garvin, McGee, Smoyer-Tomic, & Aubynn, 2009), the associational processes that normally legitimise or de-legitimise the operations of various instruments of natural resource governance have received inadequate attention. Based on an in-depth study of Ghana, this article seeks to address this imbalance by offering an account of the dynamic of NGO participation in the Extractive Industries Transparency Initiative (EITI) and its resonance with the material politics underpinning the complex operations of transparency norms in the extractive sector. It employs an overall comparative political economy approach and a heuristic model of institutional complexity, attributed to Meijer (2013), to explore NGO responses to such transparency initiatives.

The EITI was introduced in 2003 as a global standard based on transparent auditing of government company payments and receipts from resource extraction at the Lancaster Conference in the UK. Since then, the Initiative has attracted considerable research focusing on the motives and benefits that accrue to those who comply with it. However, we know very little about NGO compliance and the deep-seated contestations within the broad spectrum of civil society activism underpinning the EITI. Similarly, while Ghana's EITI compliance has received attention, the polarisations among domestic NGO activities and deep-seated apprehensions around the EITI have escaped significant scrutiny. By recasting the analytic problem of the EITI around the optic of NGO contestation and representation, the article offers a more nuanced engagement with the material politics of negotiability, contestation, and representation that fuel EITI compliance, non-compliance, and de-compliance. The overriding argument is that the understated contestations within the NGO community around the EITI, and the disparity between its platforms and representational processes of democratic accountability, ranks among the most formidable threats to the EITI's noble ambition of securing optimum resource-led transformation in implementing countries.

The empirical information used in this article was compiled from fieldwork in Ghana in 2013. Data were extracted from a variety of sources, including, documents, unobtrusive observation, and semi-structured interviews with over 100 respondents within the NGO community, government officials, parliamentarians, the Ghana Extractive Industry Transparency Initiative (GHEITI), foreign and domestic oil and mining firms, researchers and representatives of donor agencies. Legislative and policy documents, various GHEITI reports, newsletters, EITI aggregation/audit reports, parliamentary debates, and archives were also scrutinised.

The next section spells out the analytical levers based on scrutiny of the complex issues and contestations that underpin attempts at institutionalising transparency norms in the extractive sector. This is followed by an exploration of the structural undertones of NGO participation and contestations in Ghana's EITI. The concluding section examines the implications of the analysis for interrogating the EITI and similar governance instruments in the extractive sector. As used by Lang (2012), the term NGO is employed to refer to:

> any voluntary not-for-profit organisation that is bound legally to be non-political but can engage in non-institutional politics, that generates normative claims about a common

> good, and that acts on these claims as a public expert in variously publicly scaled civic spaces. (Lang, 2012, p. 13)

The notion of 'non-institutional' politics underlines the measure of fluidity that has come to characterise how NGOs are framed: between legal and policy mechanisms that often bind NGOs to the 'non-political' realm – outside and beyond the representative institutions of the political system of nation states – and the realities of advocacy and policy activism that often insert NGOs into routine politics (Beck, 1996). Moreover, while acknowledging differences in usage, the article shares the view, held by many social scientists, rejecting the presentation of NGOs as altruistic conduits for the deployment of development services, unfettered by politics.

Transparency: extractive industry dynamics

Although the roots of transparency, as a central principle of public management, and for democratic and corporate accountability, dates from the eighteenth century, the turn of the twenty-first century witnessed a phenomenal upswing in its invocation as an all-purpose recipe for the pursuit of development, broadly defined (Hood, 2007). While many definitions of transparency abound, the term often denotes a generalised acceptance for the availability of information and a disdain for secrecy in public affairs (Florini, 2010). Hence, while some commentators have raised questions about the unbridled reliance on transparency (Finel & Lord, 1999), as observed by Hood (2007), it has become one of those banal ideas that are taken as unexceptionable in discussions of governance and public management.

Transparency in the extractive sector has been sustained by an expressed desire among activists to shed light on patronage activities and sub-optimal behaviours that tend to flourish under the cover of secrecy. Its continued deployment also resonates among academics who posit that an inverse relationship exists between resource endowment and optimum developmental outcome, which can only be mitigated by norms, rules, and institutions promoting incentives for optimum behaviour. As demonstrated by various global and local instruments for extractive sector governance, the broader tenets of this policy canon are arguably informed by the need to secure order and efficiency in the presumably unstable and perverse environment around resource extraction. Analysts, mostly drawing from interdisciplinary tools of research, have paid attention to how the normative and institutional structures underpinning transparency evolved in the extractive sector (Gillies, 2010; Haufler, 2010). Another area of immense scrutiny relates to impact studies on various actors and resource-rich countries (Bracking, 2009; McGee & Gaventa, 2011). An insightful observation that has emerged from this field, which remains largely undertheorised, relates to the limits of assumptions that project the extractive sector as spaces where technocratic solutions take root irrespective of social bargains and contestations. Consequently, a theoretically informed approach to how transparency initiatives take shape in the governance landscape of resources should address how they are constructed through social and political processes. Different lines of institutional relations (between government and the public), information exchange (speed, accessibility, ease of use), and domains of transparency must also be considered (Meijer, 2013).

The rationale for paying attention to the complexity of transparency in the extractive sector can be justified on three main grounds. Firstly, at the sectoral level, an important dimension of natural resources, often linked to their territorialisation (Le Billon, 2005), is that they generate more territorial stakes than many other sectors, which usually lead to contentions over issues such as boundary delimitation, local representation, and commercial interests. As shown by Watts (2012), these points of contestation lead to new sets of governance relationship between actors at multiple scales that eventually coalesce around a 'regime of accumulation and a mode of regulation' (Hickey & Izama, 2016, pp. 5–6). The analytic objective must therefore be expanded from questions about the potential of transparency and accountability initiatives to succeed, towards an assessment of their content, the power effects they generate, and openings they offer for more contestations and negotiation (Watts, 2004).

Secondly, at a more generic level, in line with the undercurrents of the liberal democratic order, attempts at institutionalising transparency require charting a middle ground between the 'overt' and the 'secret'. To paraphrase Hood (2007, p. 195), the idea of transparency involves 'direct' observation by citizens, as in freedom of information laws enabling citizens to obtain information from government and public bodies. However, the transparency regime comes with caveats for secrecy; applying to office holders but not the public at large, and to their public activity rather than their public lives. As demonstrated by struggles between secrecy clauses that often underpin contracting processes and the continuous push by civil society activists for more information disclosure, the dynamic of steering between different spaces for openness and secrecy is less susceptible to technocratic prescriptions.

Thirdly, transparency holds differential instrumental value for actors within government, civil society, and business (Heald, 2006). Like expectations of revealed information to guide public action and oversight of resources, this instrumental value could have positive effects on natural resource governance. However, the negative impact has been well documented, especially the tendency to empower social and political elites who often manipulate information disclosure to enhance their leverage over natural resources (Florini, 2010).

The EITI: model vs. mirage

The complex dynamics of transparency, outlined in the previous section, have been reinforced by the key debates around the EITI. The initiative, which was formally unveiled by the British government at the Lancaster Conference in 2003 encapsulates three main procedures. Firstly, implementing countries ensure that all relevant companies declare to the government payments for extracting natural resources, usually through persuasion and voluntary agreements and, in a few cases, legal regulations. Secondly, these figures are compared against the government's own declaration of revenues received. Thirdly, after the data have been checked and verified by an independent body – usually an auditing firm and a multi-stakeholder committee – they are made public based on recommendations by the EITI International Board. The principal quality assurance mechanism offered by the EITI International Secretariat comes in the form of Validation, which involves an 'impartial assessment of whether EITI implementation is consistent with the EITI Standard' (EITI International Secretariat, 2014). It is through the validation mechanism that the EITI designates a country as either a 'candidate' or 'compliant'.

Since its introduction, the EITI has divided opinion. On the one hand, leading policy-makers such as Blair (2003), donors, companies, and activists have professed abiding faith in its prospects for securing 'win-win' outcomes for identifiable stakeholders in the extractive sector. Critics like Bracking (2009), on the other hand, maintain that reducing the complex value chains connecting resource extraction with sustainable development to transparent auditing and the procedural niceties of tripartite deliberation render the EITI a weak instrument. Since 2013, the EITI has undergone several reforms that have sought to ground its implementation within the extended value chain of the extractive sector. Nonetheless, as demonstrated by the case of Ghana presented below, these efforts appear to have done little to ease fears about its limitations.

Among the most damning criticisms, the EITI is faulted for its propensity to depoliticise broad-based participation and empower social elites, an observation demonstrated poignantly by its approach to civil society participation (Smith, Shepherd, & Dorward, 2012). Indeed, a central logic underpinning the EITI framework is that sustained information disclosure can garner public enthusiasm, which will, in turn, undermine perverse behaviour. Specifically, Principle 4 of the EITI Principles states that 'public understanding of government revenues and expenditure over time could help public debate and inform choice of appropriate and realistic options for sustainable development' (Extractive Industries Transparency Initiative, 2016, p. 10). As outlined by the Protocol on Participation of Civil Society, civil society is envisaged as the vehicle for generating this 'public understanding'. However, civil society participation is normally monitored through a limited lens of organisations that are 'substantively involved in the EITI process' (EITI, 2016, p. 41). To borrow from Brysk (2000, p.156), by focusing on 'substantive participants', the EITI examines 'the autonomy – but not the representativeness – of civic groups'. This has meant that one cannot rely on the EITI for a compelling account of how ordinary social agents perceive and take advantage of the shifting opportunities and constraints created by its disclosures.

The discussion so far implies that, rather than judging the EITI solely as a technocratic project that is rooted in pre-defined frameworks of social engineering, we need a more nuanced approach that probes how it webs into the intricate politics of contestations. Some of the heuristic tools that have proven resourceful in steering away from a mere interest in the degree of transparency offered by the EITI towards its dynamics is offered by Meijer (2013), who reflected on transparency from three different, but interrelated perspectives: strategic, cognitive, and institutional.

The strategic dimension relates to complexities associated with the nature of the power game and focuses on how transparency comes to be constructed, and how it impinges on the relative power of identifiable internal and external actors. While noting its contested dimensions, power is conceived here in terms of instrumental power, as a function of an actor's resources and capabilities (Bloomfield, 2017, p. 32), and involves decisions and non-decisions in pursuit of specified interests (Guzzini, 2013). It calls for attention to the variety of actors, which in our case includes NGOs, government officials, and extractive sector firms, focusing on their relative power in the socio-political construction of transparency and the ways in which changes linked to the EITI influence their power.

The cognitive dimension emphasises the importance of cognitive frames of various actors in the construction of transparency. Whereas framing theory evokes multiple perspectives, its usage often denotes the 'process by which people develop a particular conceptualisation of an issue or reorient their thinking about an issue' (Chong & Druckman, 2007, p. 104). As used by Meijer (2013, p. 432), cognitive analysis entails uncovering the meanings that actors assign to transparency and responds to questions that include: Is transparency framed as a basis for political participation, a human right, or a threat to indirect democracy? Does transparency result in the redefinition of citizens as active contributors to policies and more engaged political actors? How are the roles of government actors redefined?

The institutional analytical layer emanates from the uncertainties and contestations that emerge over the formal rules and procedures, which stipulate what types of interaction are open to democratic scrutiny. It defines the role of various actors and sets out the contours of decision making.

Whereas heuristic tools constitute gambits of imagination that do not necessarily proximate reality (Abbott, 2004), reflecting on the strategic, cognitive, and institutional dynamics offers a useful outlet to problematise some of the socio-political undercurrents of transparency. As shown in the case of Ghana, the more strategic dimension of the contestations around the EITI relates to disagreements among domestic NGOs over its value in building negotiating capital against the Government's actors and extractive companies. The cognitive framing issues identified by the study tend to converge around the endeavour of various domestic NGO actors to position themselves against attempts by the state and external donors to hegemonise the landscape of extractive sector governance. However, the article underscores incongruities in cognitive framing around individual organisational goals and perceived notions of distributional justice. The paradox associated with the institutionalisation of advocacy, which dictates the shape of NGO participation in the EITI, is also premised on questions about enforcement: Who wields (or should wield) control over policy processes? Who gets to be represented on tripartite committees, and how should they liaise with different constituencies? What are the EITI's linkages with the broader processes of democratic accountability?

Context: Ghana's path towards EITI compliance

Ghana's participation in the EITI began in 2003 following an announcement by its Government at the Lancaster House Conference that the initiative was being adopted for the mining sector. The Government's decision was part of broader reforms that were being pursued to address the governance challenges of the extractive sector, comprising gold, diamond, bauxite, manganese, and, since 2007, oil. While the sector witnessed some progress evidenced by a boost in investments, it suffered from significant governance malaise, characterised by slow and uneven progress with policy reform, the absence of sustainable management practices, distorted incentive structures, a lack of transparency, poor governance and the declining importance of non-productive environmental concerns on the part of Government (World Bank-IEG, 2014). On 17 June 2004, the Government officially launched GHEITI in Tarkwa, a mining town in the Western region. Following some notable inroads, in terms of implementation, the EITI was later extended to the petroleum sector.

Within the global EITI community, Ghana's stellar performances have not gone unnoticed: the country is credited with various country-level innovations regarding sub-national auditing as well as disaggregated and process data reporting. Many EITI-linked institutional reforms have taken place leading to an upward revision of fiscal terms associated with mining, opportunities for civil society inputs, and enhanced inter-sectoral synergies (Manteaw, 2010). Despite notable gains, Ghana's EITI has not escaped recrimination for its narrow scope, failure to generate much-needed changes in lop-sided developmental outcomes, and a propensity for NGO elitism. These criticisms were more pronounced within the NGO community given widespread claims of weak consultation and feedback from EITI representatives. Indeed, civil society participation, which takes place within the confines of the Publish What You Pay Coalition (PWYP) – a 20-member coalition of civil society organisations that is part of a global coalition of NGOs campaigning for open and accountable transparency – has featured prominently in various GHEITI validations (e.g. IDL Group & Synergy Global, 2010, p. 12). Whereas the Government's commitment to civil society participation has never been questioned, a study of 21 civil society organisations by Sakyi and Oritsejafor (2015) revealed that awareness of the EITI, especially among NGOs outside the PWYP umbrella, was low. This lack of awareness may be due to apathy rather than the usual charge of NGO elitism. Nonetheless, contestations within the NGO community for deepened representation and criticisms of the EITI's worthiness as an advocacy instrument tend to be glossed over. These contestations remain markedly different from the violent confrontations that often accompany social struggles against resource extraction, and which attract regular attention, raising questions about how the EITI concurrently opens the frontiers of transparency and contestation that shape the governance landscape of the extractive sector.

A question of power

The power dynamics of NGO contestations around EITI in Ghana are typified by different subjective judgements made by organisations and activists about its strategic value in building what Whitfield and Fraser (2009) have described as negotiating capital. This concept denotes the leverage that a party can derive from structural conditions: the dynamic global and national economic, political, ideological, and institutional context within which actors define their preferences and select their strategies. Historically, differences in the perception of negotiating capital among NGOs in Ghana date back to the civic activism that greeted neoliberal reforms in the mining sector, which were unveiled as part of the Economic Reform Programmes (ERPs) in the 1980s. The reforms attracted criticism for neglecting questions of environmental costs, threats to sustainable livelihoods, and poverty. Hence, mining communities quickly became agitating sites as various groups resorted to sporadic violence and other forms of civic action that swayed public opinion against mining firms.

NGO engagements in the extractive sector, which took centre-stage at the turn of the twenty-first century, with organisations like the Wassa Amenfi Association of Communities Affected by Mining (WACAM), the Third World Network (TWN-Ghana) and the Integrated Social Development Centre (ISODEC) were meant to build solidarity with community activists and pursue policy reform. WACAM and TWN-Ghana, for instance, initiated some high-profile judicial action over issues ranging from police brutality against protestors to environmental pollution by mining firms. NGO activism also took the form of collaborative platforms, among them the Ghana Association of Private Voluntary Organizations in Development (GAPVOD) and the National Coalition on Mining. While GAPVOD's antecedent predates the ERPs era, its engagements in tax reforms and public financing under the auspices of ISODEC spilled over into the mining sector, culminating in the launch of the local chapter of the PWYP campaign. The Coalition was launched in 2001 by TWN-Ghana and some NGOs with the goal of working 'for the respect of human rights in mining and optimal national and citizens' benefits from the country's mineral resources' (National Coalition on Mining, 2014, p. 1). These unilateral and collective efforts by NGOs, coupled with sporadic violent agitations, brought enormous pressure on the government and mining companies. In response, the Government and mining firms began disclosing mining sector revenues through specific designation in budget statements and press releases from the Chamber of Mines. The EITI was therefore envisaged partly to institutionalise the evolving transparency practices and manage the tension around mining. A representative of the Chamber, for instance, disclosed that it was the mining companies who 'actually pushed' the Government to adopt the EITI because, for them, 'the Initiative is like getting a third-party person to tell a good story about you' (interview, 2013).

Nonetheless, its unveiling polarised NGOs along two mainlines. Firstly, for members of the Coalition on Mining, the EITI was a 'ploy' to diffuse what its members considered as a growing momentum for activism in the mining sector (interview with NGO activist, 2013). Representatives from this group maintained that the EITI failed to address the 'core problems' of the extractive sector, which they identified as lax taxation, human rights abuses, environmental degradation, and social dislocation (interviews with TWN activists, 2013). They also suggested that sustained market reforms in the sector had eroded their negotiating capital vis-à-vis the Government and companies; a phenomenon they argued can only be reversed by rectifying this imbalance and forging deeper networks with broader civil society. Their sentiments resonated with a general fear that is often associated with the tendency of NGO-government/firm interdependence, to tame radical capital campaign strategies in the mining sector (Bloomfield, 2017).

The flipside of concerns about taming activism, as noted by Bloomfield (2017), is that NGOs that partake in such collaborations may secure leverage by gaining insider knowledge and influence. This was the position echoed strongly by NGOs under the PWYP. Although they accepted its limited scope, they maintained that the EITI provided an opportunity for enhanced advocacy in the extractive sector. They argued that, despite its limitations, the Global Sourcebook, which espoused the main EITI principles, encouraged countries to be ambitious and apply in-country variations.

Cognitive framing

Differences in cognitive frames also played a significant role in NGO participation in the EITI. Even actors who participate in the EITI, as elsewhere, often share diverse expectations that are hardly met. For the politics of transparency, this mostly relates to compromises between the overt and secret. As Andrews (2016) discovered in Ghana, the clamouring for the continued expansion of the frontiers of public disclosure, followed by far-reaching reforms often meets resistance from the Government and extractive companies.

Among the NGO community, cognitive frames tend to converge on various issues. At its inception, both critics and proponents argued that the EITI framework was not new, given the practices of disclosures that pre-dated it in Ghana. This broad-scale consensus led NGOs to push for the expansion of the EITI. The aggregator was, for instance, tasked with conducting a 'process and financial audit', namely analysis of 'the historical documentation on production, exports and payment of royalties for the minerals produced in the country' (Boas & Associates, 2007, p. 19). NGOs continued to push for the expansion of the landscape of transparency from the focus on significant producers – major mining and petroleum companies – who are the focus of EITI reports, to small-scale contractors and service providers, while placing significant weight on contract transparency.

NGO activists also agreed that EITI implementation should rest on 'home-grown' structures, although the Initiative is considered as an international norm. This position came to light when they rallied around the Government in a dispute with donor agencies over the appointment of Boas & Associates, a domestic auditing firm, to undertake EITI aggregation. In a dispute that reflected the diversity of expectations associated with the EITI at the level of Government–donor relations, the World Bank and the international EITI Secretariat raised concerns about the appointment of the local auditor and insisted that the Government should employ an Aggregator from among seven recommended international consultants as a precondition for disbursements for GHEITI. When the Government, together with the newly established GHEITI Steering Committee, insisted on maintaining Boas & Associates, the Bank refused to fund the first GHEITI aggregation (IDL Group & Synergy Global, 2010, p. 20).

Nonetheless, NGO's cognitive frames tend to diverge. The more apparent dimensions of these variations relate to ideological goals and mission statements. For instance, respondents from TWN-Ghana intimated that using global and regional coalitions, rather than the more institutionalised EITI, remains a core aspect of their mission. In an interview, a labour leader in the petroleum sector also underscored a measure of misalignment between his union's organisational goals and the EITI's activities as partly responsible for their limited participation. Beneath these ideological differences, however, the most profound cognitive incongruities emanated from normative judgements about what a just outcome(s) from resource distribution should be. To develop this argument, it is important to revisit the EITI's claims about 'win-win' outcomes for compliant stakeholders. These outcomes, as outlined by the EITI's Oslo-based office, include an improved investment climate, public accountability and good governance, and prevention of resource-based conflicts (EITI International Secretariat, 2015). The normative dimensions of these EITI fruits are rooted in a relatively libertarian conception of distributive justice that rationalises private appropriation of natural resources, only if agents compensate for the competitive value of the rights that they claim over the specified resource (Vallentyne, 2007, p. 559). However, as pointed out by Vallentyne (2007) and Pogge (1994), while this conception supports the unilateral appropriation of shares of resources, it is insufficiently sensitive to distributive considerations, especially in a context where greater disparity in wealth and opportunities for wellbeing prevail. Consequently, some critics have dismissed the prospects for a 'win-win' outcome and underscored the need to take a 'zero-sum' view, resolved in favour of communities that hold proprietary rights to extractive resources (Bourgouin & Haarstad, 2013). These contrasting precepts of distributive justice surfaced in interactions with various activists. A gulf exists between what the Initiative offers and what the public perceive as fair outcomes from resource extraction. For instance, a member of GHEITI's Steering Committee observed:

> From our dissemination workshops, the feedback is not only an interest in whether we got revenue or not, but how the companies would address the legacy issues; the environmental pollution; the water stress in communities because of mining; the takeover of community lands. (Interview with NGO activist, 2013)

Community activists in the Western Region, like Friends of the Nation, who were also interviewed, indicated that the revenue-centeredness of GHEITI has masked other aspects of the extractive sector, specifically questions about environmental damages and costs, as well as threats to livelihoods. Within these contrasting notions of distributional outcomes, various NGOs have sought to appropriate the discourse of 'progressive justice' from resource extraction by positioning within and outside the EITI. NGO activists on the EITI's platforms maintained that they have managed to secure some 'wins' to address the lop-sided distributional outcomes from the extractive sector. Conversely, their critics maintained that boycotting the EITI enabled them to side with the developmental dimensions of distributive justice, which the EITI does not offer.

The effects of these choices underscore the limits to claims of aloofness from politics that are often voiced against activists who boycott such deliberative platforms of the extractive sector. As Bloomfield (2017) has observed elsewhere, activist pressure outside the confines of collaborative platforms could induce cumulative effects in the mining sector. Consequently, NGOs who boycott the EITI tend to lay claim to a measure of ideological purity aligned with the progressive dimensions of distributional justice in a manner less akin to activists on EITI platforms. However, as shown below, NGO opponents to the EITI suffer a great deal of disadvantage that emanates from a broader paradox related to institutional politics.

Unbinding the rules of the game

The institutional dynamic of NGO participation in Ghana's EITI resonates strongly with what Lang (2012, p. 8) described elsewhere as the 'paradox of NGOisation': the increasing institutionalisation of advocacy. This paradox relates to everyday bargains that NGOs often strike between the pursuit of institutional effectiveness and public advocacy. According to Lang (2012, p. 8), 'as NGOs become stronger institutional players and are welcome into civic dialogues with local, national, or transnational political institutions, the incentives to strategically limit public advocacy increase'. Equally, NGOs who decide to stay on the side-lines of institutional political processes tend to fuel a perception of 'aloofness from politics', which could inadvertently jeopardise institutional leverage (Lang, 2012, p. 8). The overriding fallout of this paradox is that many NGOs have become experts in institutional advocacy and lobbying at the expense of generating broader public debates.

In the extractive sector, the drive for NGOisation is the widespread perception that erroneously presents NGOs as the legitimate interlocutors of the public, whose presence on tripartite institutions represents the inclusion of marginalised voices. However, the failings of NGO participation in such tripartite arrangements, in the shape of elite manipulation and co-optation, as well as detachment from wider stakeholder communities have been widely reported (Edwards & Hulme, 1992; Jenkins, 2001; Oppong, 2016). Similarly, in various interviews, critics of GHEITI observed that, given the unwillingness of governments to disclose information, due to what they considered as a tendency for secrecy in governmental affairs and contractual obligations, the choice of Government as the host of EITI renders it a less plausible instrument. To quote another respondent:

> [We] didn't join the EITI because we are concerned about how it is structured; the very entity which is not putting information out there, the state, is the one hosting the EITI and now even within the EITI, people are complaining. (Interview with TWN official, 2013)

Indeed, concerns about whether the Government, as an entity, could be trusted to implement EITI reforms are not uncommon. A wealth of studies about the EITI has singled out the public as the silent partner, given the general tendency of governments to exclude broad sections of civil society (Aaronson, 2011). However, a leading architect of Ghana's EITI observed that, while Government and companies tend to exercise much power, 'it is more of a perception than the reality' in Ghana (interview with NGO activist, 2013). His observation reverberated with other members of GHEITI's Steering Committee who argued that, because they take decisions by consensus rather than voting, the dominance of governmental actors, which amounted to 15 (out of 22) during the study, is less decisive. The institutional quagmire that this position evokes is that GHEITI survives more on the goodwill of dominant stakeholders rather than through more formalised processes. This problem has been reinforced by the absence of any binding legislation that could clarify enforcement mechanisms of the EITI and strengthen linkages with the Ghanaian parliament. Parliamentarians who were interviewed intimated that, although informal interactions with GHEITI take place through workshops and briefings, none of the eight GHEITI reports was debated in parliament or scrutinised by the Public Accounts Committee (PAC), Mines and Energy Committee, and the Finance Committee. They explained that neither the PAC, which is mandated to scrutinise reports emanating solely from the Auditor-General's department, nor other parliamentary committees had a clear mandate to follow up on GHEITI reports (interviews with Members of Parliament, 2013).

Other institutional weaknesses of GHEITI result from the poor connections with broader constituencies that are represented on its platforms. Several NGOs, within and outside the PWYP-Ghana coalition, in contrast to the claims by their representatives, complained about the limited institutionalised feedback. An informant from the Ghana Integrity Initiative went a step further when he averred that this poor feedback is because 'it is not civil society that sent[them] there'. While various dissemination workshops have shown promise in building the linkages with the public, the weak feedback with broader civil society has undermined the EITI's legitimacy. This problem points to a wider trend in NGOisation paradox: NGOs who boycotted the EITI admitted to a growing marginalisation from the frontiers of institutional politics in the extractive sector (interviews with NGO activists, 2013). While they attributed this issue to donor and governmental activities that tend to focus on the EITI, it underscores the pitfalls of aloofness from institutional politics. In contrast, those who participate in EITI maintained that it has offered an invaluable opportunity for institutional leverage and public advocacy. Nonetheless, by identifying with the *status quo,* a growing disjuncture has opened between EITI representatives and other activists who claim to be on the ground.

Conclusion

There is no doubt the transparency agenda in the extractive sector represents a progressive step. By institutionalising transparency through purposive information disclosure and tripartite disclosure, the EITI and similar governance instruments seeking to boost the prospects for resource-led transformation have proven pivotal in expanding the landscape for civil society engagement. Nonetheless, a scrutiny of the dynamics of NGO participation and the subjective judgements that various social actors make about the strategic, cognitive, and institutional subtleties of transparency has underscored some limits of the transparency agenda, especially its reductionism in resource-rich countries. Far from the unidirectional view that upholds the diffusion of pre-selected norms of transparency, the presentation in this article mirrors the understated contestations and bargains associated with natural resources, which shape the trajectory of transparency initiatives. In many respects, this tension emanates from what Pogge (2002) characterised elsewhere as the intricate 'privileges' accorded governments based on territorial sovereignty over natural resources, irrespective of normative orientations and wider distributional contestations and outcomes. Consequently, attempts at filtering norms of appropriate resource management into the national landscape often come up against this impenetrable wall of privileges.

The EITI's novelty lies in placing governments at the epicentre of information disclosure while securing policy access for NGO representatives through multi-stakeholder deliberation. As demonstrated by countries like Ghana and Nigeria, the openings afforded by these EITI platforms have been useful not only at drawing public attention to revenue transparency but also at securing some policy and institutional reforms (Manteaw, 2010). However, the insights offered by deeper examination of NGO contestations in this article underscore a growing disjuncture between various EITI platforms and representational processes of accountability that tends to attract less scrutiny. To avoid criticism for promoting transparency without a corresponding improvement in accountability, the EITI needs to attain a closer fit with the strategic, normative, and institutional values that different organisations attach to it in implementing countries.

Notes on contributor

Nelson Oppong is a lecturer in international development at the University of Bath and member of the Editorial Working Group of the *Review of African Political Economy*. His research scrutinises the intersections between global norms and domestic drivers of reform in resource-rich countries. With over 10 years of field research and policy work as a consultant in various international organisations, he also maintains active interest in the unique trajectories of governance and institutional development in small island developing states and low-income countries across Africa, the Caribbean, and the Pacific.

References

Aaronson, S. A. (2011). Limited partnership: Business, government, civil society, and the public in the Extractive Industries Transparency Initiative (EITI). *Public Administration and Development*, *31*(1), 50–63. doi:10.1002/pad.588

Abbott, A. (2004). *Methods of discovery: Heuristics for the social sciences*. New York: WW Norton & Company.

Agrawal, A., & Gibson, C. C. (1999). Enchantment and disenchantment: The role of community in natural resource conservation. *World Development, 27*(4), 629–649.

Andrews, N. (2016). A Swiss-Army knife? A critical assessment of the Extractive Industries Transparency Initiative (EITI) in Ghana. *Business and Society Review, 121*(1), 59–83. doi:10.1111/basr.12081

Beck, U. (1996). World risk society as cosmopolitan society? Ecological questions in a framework of manufactured uncertainties. *Theory, Culture & Society, 13*(4), 1–32.

Blair, T. (2003, June 17). *Speech at the Extractive Industries Transparency Initiative (EITI) London conference.* Retrieved from http://collections.europarchive.org/tna/20070701080507/http://www.dfid.gov.uk/pubs/files/eitidraftreportspeech.pdf

Bloomfield, M. J. (2017). *Dirty gold: How activism transformed the jewellery industry*. Cambridge, MA: The MIT Press.

Boas & Associates. (2007). *Report on the aggregation/reconciliation of mining benefits in Ghana January–June 2004*. Accra: Author.

Bourgouin, F., & Haarstad, H. (2013). From 'good governance' to the contextual politics of extractive regime change. In J. Singh & F. Bourgouin (Eds.), *Resource governance and developmental states in the global south: Critical international political economy perspectives* (pp. 87–106). Basingstoke: Palgrave Macmillan.

Bracking, S. (2009). *Hiding conflict over industry returns: A stakeholder analysis of the Extractive Industries Transparency Initiative.* Manchester: Brooks World Poverty Institute.

Brysk, A. (2000). Democratizing civil society in Latin America. *Journal of Democracy, 11*(3), 151–165. doi:10.1353/jod.2000.0049

Chong, D., & Druckman, J. N. (2007). Framing theory. *Annual Review of Political Science, 10*, 103–126. doi:10.1146/annurev.polisci.10.072805.103054

Edwards, M., & Hulme, D. (1992). Scaling up NGO impact on development: Learning from experience. *Development in Practice, 2*(2), 77–91. doi:10.1080/096145249100076691

EITI International Secretariat. (2014). *The EITI glossary*. Oslo: EITI.

EITI International Secretariat. (2015). *Benefits from implementing EITI*. Oslo: EITI.

Extractive Industries Transparency Initiative. (2016). *The EITI standard 2016*. Oslo: Author.

Finel, B. I., & Lord, K. M. (1999). The surprising logic of transparency. *International Studies Quarterly*, 315–339. doi:10.1111/0020-8833.00122

Florini, A. (2010). The national context for transparency-based global environmental governance. *Global Environmental Politics, 10*(3), 120–131. doi:10.1162/GLEP_a_00017

Garvin, T., McGee, T. K., Smoyer-Tomic, K., & Aubynn, E. A. (2009). Community–company relations in gold mining in Ghana. *Journal of Environmental Management, 90*(1), 571–586. doi:10.1016/j.jenvman.2007.12.014

Gillies, A. (2010). Reputational concerns and the emergence of oil sector transparency as an international norm. *International Studies Quarterly, 54*(1), 103–126. doi:10.1111/j.1468-2478.2009.00579.x

Guzzini, S. (2013). *Power, realism and constructivism*. London: Routledge.

Haufler, V. (2010). Disclosure as governance: The Extractive Industries Transparency Initiative and resource management in the developing world. *Global Environmental Politics, 10*(3), 53–73. doi:10.1162/GLEP_a_00014

Heald, D. (2006). Transparency as an instrumental value. In C. Hood & D. Heald (Eds.), *Transparency: The key to better governance?* (pp. 59–73). Oxford: Oxford University Press.

Hickey, S., & Izama, A. (2016). The politics of governing oil in Uganda: Going against the grain? *African Affairs, 116*(463), 163–185.

Hood, C. (2007). What happens when transparency meets blame-avoidance? *Public Management Review, 9*(2), 191–210.

IDL Group, & Synergy Global. (2010). *Validation of the Extractive Industries Transparency Initiative in Ghana. Final report*. Accra: GHEITI.

Jenkins, R. (2001). Mistaking 'governance' for 'politics': Foreign aid, democracy and the construction of civil society. In S. Kaviraj & S. Khilnani (Eds.), *Civil society: History and possibilities* (pp. 250–268). Cambridge: Cambridge University Press.

Lang, S. (2012). *NGOs, civil society, and the public sphere.* Cambridge: Cambridge University Press.
Le Billon, P. (2005). The geopolitical economy of 'resource wars'. In P. Le Billon (Ed.), *The geopolitics of resource wars: Resource dependence, governance and violence* (pp. 1–28). Oxford: Frank Cass.
Manteaw, S. (2010). Ghana's EITI: Lessons from 96 mining and policy implications for oil. *Ghana Policy Journal. Special issue: Ghana's Petroleum industry: The prospects and potential impediments towards good governance standard*, 96–109.
McGee, R., & Gaventa, J. (2011). Shifting power? Assessing the impact of transparency and accountability initiatives. *IDS Working Papers, 2011*(383), 1–39.
Meijer, A. (2013). Understanding the complex dynamics of transparency. *Public Administration Review, 73*(3), 429–439. doi:10.1111/puar.12032
National Coalition on Mining. (2014). *National coalition on mining calls on President Mahama to rescind Dr Aubynn's appointment as head of Minerals Commission.* Accra: Third World Network.
O'Higgins, E. R. E. (2006). Corruption, underdevelopment, and extractive resource industries: Addressing the vicious cycle. *Business Ethics Quarterly, 16*(2), 235–254. doi:10.1017/S1052150X00012823
Oppong, N. (2016). Ghana's public interest and accountability committee: An elusive quest for 'home-grown' transformation in the oil industry. *Journal of Energy & Natural Resources Law*, 1–24. doi:10.1080/02646811.2016.1179464
Pogge, T. (2002). *World poverty and human rights: Cosmopolitan responsibilities and reforms.* Cambridge: Polity Press.
Pogge, T. W. (1994). An egalitarian law of peoples. *Philosophy & Public Affairs, 23*(3), 195–224. doi:10.1111/j.1088-4963.1994.tb00011.x
Sakyi, E. K., & Oritsejafor, E. O. (2015). Extractive sector policymaking and governance in Ghana: A study of the role, knowledge and capacity challenges of civil society organizations. *Journal of Public Management & Social Policy, 21*(1), 4.
Smith, S., Shepherd, D., & Dorward, P. T. (2012). Perspectives on community representation within the Extractive Industries Transparency Initiative: Experiences from south-east Madagascar. *Resources Policy, 37*(2), 241–250. doi:10.1016/j.resourpol.2011.01.001
Vallentyne, P. (2007). Distributive justice. In R. Goodin, P. Pettit, & T. Pogge (Eds.), *A companion to contemporary political philosophy* (2nd ed., pp. 549–562). Oxford: Wiley-Blackwell.
Watts, M. (2004). Resource curse? Governmentality, oil and power in the Niger Delta, Nigeria. *Geopolitics, 9*(1), 50–80. doi:10.1080/14650040412331307832
Watts, M. (2012). A tale of two gulfs: Life, death, and dispossession along two oil frontiers. *American Quarterly, 64*(3), 437–467. doi:10.1353/aq.2012.0039
Whitfield, L., & Fraser, A. (2009). Introduction: Aid and sovereignty. In L. Whitfield & A. Fraser (Eds.), *The politics of aid electronic resource. African strategies for dealing with donors* (pp. 1–25). Oxford: Oxford University Press.
World Bank-IEG. (2014). *Project performance assessment report Ghana. Natural resources and environmental governance first development policy operation (IDA-44260).* Washington, DC: World Bank.

Gendered participation in community forest governance in India

Satyapriya Rout

ABSTRACT

Recent forest governance practices in India have responded to environmental change and subsequent livelihood insecurity by focusing resource governance policies on communities. A paradigm shift has occurred involving participatory inclusive bottom-up approaches, rather than state-centric, top-down forestry. With the formulation of the 1988 National Forest Policy, several variants of participatory models of forest governance – social forestry, community forestry, joint forest management – have been tried out, with differing degrees of success. The 2006 Forests Right Act adopts a rights-based approach to participatory forestry to address the serious concerns of environmental degradation, livelihood insecurity, tenure reforms and questions of autonomy and identity of forest-dependent communities. Using mainly qualitative methodology, this paper reviews forest governance policies and undertakes a critical examination of recent participatory forestry practices. Drawing empirical evidence from two community-based forest governance institutions in the state of Odisha in eastern India, the paper demonstrates how participatory forestry programmes, albeit successful, may be exclusionary with regard to women's engagement in cases where their involvement is under-represented.

Introduction

Forests are an essential source of oxygen, an effective carbon sink and a valued livelihood support system in the context of developing countries. Forest governance, therefore, constitutes a core theme in debates about global climate change and environmental degradation. Since the late 1980s, forest governance practices in developing countries like India have responded to ongoing environmental change and subsequent livelihood insecurity by bringing communities to centre stage in resource governance policies to ensure sustainability of natural resources. With the promulgation of the 1988 National Forest Policy, the Indian forestry sector witnessed several variants of participatory models of forest governance with varying degrees of success. In the early twenty-first century, a notable addition to the panoply of models was the 2006 Forest Rights Act (FRA), which adopted a rights-based approach to participatory forestry in an effort to address the serious concerns of environmental degradation, livelihood insecurity, tenure reforms and questions of autonomy and identity of forest-dependent communities.

While recent developments, both in terms of policy and practice, have focused on community's engagement in forestry, there is no straightforward reason why participatory forestry should itself be termed 'inclusionary'. Drawing empirical evidence from two community-based forest governance institutions from the Indian state of Odisha, the present paper attempts to demonstrate how successful participatory forestry programmes may become 'exclusionary' due to their inadequate engagement of women. Since the study on which the paper is based attempts to analyse the implications of women's exclusion from successful participatory models, we chose two participatory forest governance institutions with different approaches to the integration of women. Such a selection allowed us to investigate the implications of women's exclusion on policy and practice in otherwise exemplary participatory models. While both the institutions in our study exhibited a sufficient degree of collective action and community participation to enable sustainable management of local natural resources, the situation was very different regarding women's involvement. These two cases thus provide evidence that can be used to inform participatory forest policies about the contribution that women can bring to institutions of forest governance when they are involved in decision-making arrangements. The aim of the research was both to inform resource governance policy and to propose evidence-based recommendations for improving resource governance practices in developing countries like India.

Gender and environmental governance: a theoretical understanding

Environment and development discourse involves different population groups based on gender, class and ethnicity. Women are today considered as important primary stakeholders, and their role in environmental protection and resource governance has been widely accepted (Agarwal, 2010). Theoretical debate on gender and environmental governance focuses mainly on women's closeness to the immediate environment and their knowledge about natural resources and their utilisation. The ecofeminist argument, while presenting environmental destruction as a reflection of the domination and subordination of nature and women by men, celebrates the conservation potential of women due to their closeness to nature (Mellor, 2007; Miles & Siva, 1993). Ecofeminism argues that, as women are closer to nature than men, they have a greater potential for conserving resources. Women's anatomy and reproductive capacities of childbearing and rearing are associated with the affirmation of this close connection to nature (Miles & Siva, 1993). Ecofeminist argument equates men's domination of women in society with men's control and exploitation of natural resources. It attempts to establish a linkage between the ecology of nature and the biology of women's bodies; and between the exploitation of nature and that of women's wombs (Krishna, 2008).

The ecofeminist argument is often criticised for its essentialisation of women, femininity and motherhood without a proper understanding of the material conditions and social realities of women across the world (Cuomo, 1998). In confronting the universalistic and ideological construction of ecofeminism, Agarwal (2001) argues that women's responses to environmental degradation need to be contextualised and understood in accordance with the material reality that emerges from their everyday interaction with nature and dependence on it. Focusing on women's conservation potential, Agarwal (1992) maintains that, rather than a biologically rooted connection between women and nature, there exist a culturally specific ecological location, social divisions and survival needs that enhance women's affinity to the environment compared to that of men.

Agarwal's 'feminist environmentalism' approach traces the need for women's involvement to their material dependence on natural resources for survival. She argues that women, being most dependent upon forests for subsistence, are sensitive to ecologically sustainable development needs. Owing to the gender division of labour in developing societies, women are the primary responsible members in households for fetching fuel and fodder, and are thus more likely to be adversely affected by any environmental degradation. While complying with their gender specific obligations, women gain special knowledge about nature and its regeneration through their day-to-day engagement with nature. 'Women, therefore, could be seen as both victims of the destruction of nature and as repositories of knowledge about nature, in ways distinct from men of their class' (Agarwal, 1992, p. 126).

The policy context: participatory forest governance regimes in India

Despite the significant role that rural communities play in the regeneration of natural resources, until the 1980s forest governance policies in India made no reference to community participation or the involvement of women. The first-ever forest act of India, the Indian Forest Act 1865 (revised in 1878), stipulated that the State should demarcate and establish control over valuable tracts of forest. State-controlled forest governance continued even after independence, and forest laws and policies were promulgated on the assumption that national interest was best served by capitalising on natural resources.

The forest governance policies of India started to recognise people's need to have control over forest resources only during the 1970s. Greater acknowledgement of successful cases of community-based forest governance, grassroots movements by forest communities demanding more voice in decision-making, and overall policy changes involving the local population in development and decentralisation are cited as some of the reasons for this paradigm shift in forest governance in India (Poffenberger & McGean, 1996). The new forest governance paradigm attempted to meet the needs of the population by actively involving them in the creation of forest resources in community lands. Based on the recommendations of the National Commission on Agriculture, the Social Forestry Programme was launched in 1972, focusing on participative plantation activities to meet the twin goals of increased availability of fuelwood and reduced demands on natural forests (Robinson, 1998).

Against a backdrop of increasing rates of deforestation, loss of biodiversity and emerging local struggles against their exclusion from forest governance practices, the National Forest Policy was launched in 1988, which gave primacy to environmental sustainability, soil conservation, maintenance of ecological balance and, above all, people's participation in forestry activities. On 1 June 1990, the Government of India adopted a National Joint Forest Management (JFM) Resolution, which set guidelines for partnerships between local communities and the Forest Department through Forest Protection Committees (FPCs).

Rights-based approaches to participatory forestry from the mid-1990s

Community participation in forest governance in India received a new impetus in the mid-1990s, as participation increasingly came to be seen as a 'right'. Two important policy frameworks were introduced, which have facilitated community's engagement in forest governance practices.

First, the Panchayats Extension to Scheduled Areas (PESA) Act, which was enacted in 1996, attempted to uphold the unique identity of the Scheduled Tribes (STs) and their rights over their land, forest and self-governance. The PESA Act suggested that 'every Gram Sabha [village council] should be competent to safeguard and preserve the traditions and customs of people, their cultural identity, and community resources, including land, forests and water'. Second, the Scheduled Tribes and Other Traditional Forest Dwellers (Recognition of Forest Rights) Act 2006, known as the Forest Rights Act (FRA), acknowledged that forest-dwelling communities are integral to the very survival and sustainability of the forest ecosystem. FRA recognised the rights of the local forest communities and vested in them the right to use, manage and conserve forest resources. It provided for the recognition of 13 different rights that are central to the lives and livelihoods of tribals and other forest-dwelling communities. It provides for the involvement of the forest-dwelling communities through the formation of Forest Rights Committees (FRCs) at village level.

Case studies of community forest governance

This section examines and compares the contexts in which the women's participation was observed. It explores the role of community forestry in sustaining the livelihood of the local population, and the institutional arrangements for ensuring the engagement of women in forest governance.

Contexts and methods

The empirical work for the present paper was carried out in the state of Odisha in India, where forests cover 31.06% of the total geographical territory, spreading over 48,366 square kilometres. Odisha accounts for 7.13% of the total forest cover of India, and occupies fourth position in terms of coverage. The state of Odisha was chosen because community involvement in the governance of forest resources through locally evolved institutions is widespread. Odisha also has a larger concentration of STs (22.8% in the 2011 Census) who mostly depend upon the forest for their livelihood.

For the empirical work, two FRCs created under the auspice of FRA 2006 were chosen from the Ranpur Block of Nayagarh district of Odisha. Although both were examples of participatory governance, the two institutions reported significant differences in women's participation. Women members of the community participated in equal terms in the Baghadevi Forest Protection Committee (BFPC), while in the Khatia Forest Protection Committee (KFPC), women's names were just included in the Executive Committee (EC) without any real participation and engagement. Focus group discussions and in-depth interviews were carried out with the eight office bearers of each FPCs. Ten key-informant interviews were also conducted with female members to understand women's perspectives on participatory forest governance and their inclusion/exclusion from such participatory initiatives.

Over the years, two broad types of FPCs have evolved to give shape to community forestry in India: self-initiated and state-sponsored FPCs. In self-initiated FPCs, rural communities are engaged in forest protection in line with their customary practices and traditional norms. The history of self-initiated forest protection in Odisha dates back to the early twentieth century, with evidences of voluntary forest protection by communities in Lapanga in the Sambalpur district in 1936 (Sundar, Mishra, & Peter, 1996). The state-sponsored FPCs were designed in response to specific government participatory forestry programmes, such as the Social Forestry, JFM or FRCs under FRA 2006.

Involvement in forest protection activities in our two localities dates from the early 1990s. After the implementation of the FRA, both communities applied for recognition, and were registered as FRCs under the Act. Despite their successful attempts at participation, they differed qualitatively in terms of their composition and women's engagement in forest protection. The BFPC emphasised the issue of women's participation more than the KFPC. Not only did they include women in EC, but they also ensured their active engagement in decision-making. Since both BFPC and KFPC were neighbouring localities with similar socio-economic conditions, the conditions of the forest, the nature of people's dependence on it, and their history of forest protection were very similar in nature.

Forest degradation, livelihood insecurity and the emergence of community forestry

The legal status of the forest for both study sites is described as 'reserved forest', and is, thus, under the jurisdiction of the Forest Department. The forest is composed of different varieties of tree species, with a predominance of sal (*Shorea robusta*), teak (*Tectona grandis*) and bamboo. The forest is closely related to the inhabitants' everyday lives: they depend primarily on the forest for three major purposes: food, fuelwood and fodder. While dependence upon the forest for food is more prevalent among the lower socio-economic strata, fuelwood collection is universal in the locality. Households collect edible items from the forest such as seasonal fruits, mushrooms, green leaves and leafy vegetables for their daily consumption.

The status of reserved forest requires the enforcement of certain formal regulations for access and use. However, these de jure regulations were hardly observed on the ground, making the forest a de facto open access resource (McKean, 2000). Despite their dependence on the forest, the communities in the case studies had not framed any specific norms or rules (either formal or informal) concerning access and use of forest produce. It was not until 1994–1995 that informal arrangements started to emerge to regulate access to forest resources in the locality. Increasing degradation of the forest, scarcity of forest produce for everyday use and related livelihood insecurity were reported as reasons for the initiation of protection activities in both the forest governance institutions. The increasing presence of wood mafia and smugglers in the locality, and illegal cutting of trees for private gain were reported as additional reasons for initiating community mobilisation for the protection of forests. During the early 1990s, community mobilisation was carried out with active involvement from local school teachers, village elders, youth clubs and civil society to create awareness of the benefits of forest protection. After much deliberation, both the communities formed FPCs, and designed ground rules for access and use of forest resources. After the implementation of FRA in 2006, both communities applied to the Forest Department for official recognition under the Act, and were registered as FRCs.

Institutional arrangements for participation in forest governance

Institutional arrangements for forest protection in both the committees involved an EC consisting of a president, vice-president, secretary and five more members from the community, chosen unanimously. All the adult members of the community constituted the General Body (GB) of the FPC. While the GB of both the FPCs collectively formulated the rules of forest protection, it was the responsibility of the EC to implement and monitor them, and impose sanctions for violations.

Both BFPC and KFPC adopted participatory (albeit different) methods collectively to frame the rules of forest protection. The GB of the FPCs met twice a year to deliberate about the framing and implementation of rules. According to the ground rules, all the primary users can gain access at any point of time to the forest without permission from the EC for their livelihood requirements. However, if they need to collect timber for construction purposes, they must seek permission from the EC. Outsiders were not permitted to access the forest, and if an outsider was found in the forest s/he would be brought before the EC for them to decide what action should be taken.

The EC of both FPCs met at least once a month to discuss and deliberate about implementation of the rules. The venue and timings of the meeting were announced to all so that other interested members could attend. Since communities were close-knit with face-to-face relations, the meetings were mostly informal in nature, and were conducted either in the village school building or in a community space. In the meetings, the EC decided whether to grant permission to individual households to collect timber for construction purposes; it prescribed sanctions for violation of rules and determined the times and duration of forest closure (banning of access to forests) for protection purposes. The EC also discussed and attempted to resolve conflicts concerning access and use of forest resources. Conflicts involving outsiders and wood mafia were usually reported to the Forest Department.

Women and participatory forest governance

Despite the emphasis on women's equal participation in FPC in JFM resolution of 1990 and the 2006 FRA, our empirical analysis revealed that the two FPCs in the study represented different models of participation by women. Participation was assessed using indicators such as regularity of attendance at FPC meetings, participation in discussions at planning and decision-making meetings, and the contribution of time, resources and labour to forest protection activities.

The EC of each of the FPCs was composed of eight members, with three female members in KFPC and four in BFPC. Participation in general meetings was quite high for both the FPCs. However, the female members of the KFPC very rarely attended the monthly meetings of the EC and remained absent from the decision-making process. Their absence did not mean that the institution was not participatory in nature, but rather that decision-making was male dominated. In contrast, a substantial number of women participated not only in the general meeting of BFPC but also in the discussions to decide on the rules of forest closure, regulation of access and use. In addition, they devoted time and labour to forest protection activities. They guarded the forest in groups and in rotation against forest theft and violation of the rules of protection.

Implications of exclusionary participation for policy and practice

Participation in development and resource management literature is much celebrated both as the means to achieve stated project outcomes, and as an end in itself. In the context of rural water supply projects, for example, Narayan (1995) reiterates that participation results in project effectiveness, economic efficiency, empowerment and social equity. Participation ensures citizenship rights, provides an opportunity to express voice and results in building the agency and capacity of participants (Paul, 1987; White, 1996).

From this perspective, women's absence from KFPC could be considered as an indicator of the project's failure, irrespective of the availability of resources for the community. Further, excluding women while at the same time including men in a seemingly participatory institution disproportionately affected the already existing unequal gendered power relations. Rather than being an instrument of equity and empowerment, such a model of participation potentially disempowered women and had a negative impact on issues of equity. Our study conceptualised this model of participation as 'exclusionary participation', with far-reaching implications for women, forest governance and the participatory policy framework.

The implications of exclusionary participation were examined by comparing attitudes, behaviours and interests of female members in the two FPCs. A marked difference was observed in women's attitude, their behaviour towards the forest and their interest in forest protection activities. In BFPC, women possessed a more informed and positive attitude towards forests and their protection than their counterparts in KFPC. Participation in meetings, discussions and activities of forest protection gave the BFPC women a new kind of identity, and an awareness of the multiple benefits of being surrounded by forests. The otherwise illiterate rural women of BFPC possessed working knowledge about the role of forests in arresting soil erosion, environmental degradation and climate change. By contrast, KFPC women did not exhibit any positive interest in forest protection, and could not associate themselves with the protection activities of their committee. According to one of the female members of KFPC:

> Forest protection is a men's activity in our village. They claim that they protect the forest and thus instruct us not to cut any trees or branches. But how can we do that? We cannot cook without fuelwood. Where shall we get fuelwood, if not from forest?

The KFPC women considered forest protection and the fulfilment of day-to-day requirements from the forest as conflicting with each other, and accorded greater priority to the latter. There was a clear understanding among the women in KFPC that their interests, priorities and requirements were not taken into account in the participatory model of forest protection. In contrast, BFPC women associated themselves with activities of forest protection, and possessed an informed and positive attitude. This evidence would seem to support the arguments put forward in the context of Northern America by the feminist political theorist Phillips (1995), who maintained that women's presence in politics would bring not only a quantitative change, but also a qualitative change in attitudes, behaviours and interests.

Without undermining the virtue of participation, it can be argued that 'exclusionary participation', which excludes women from participatory processes, has potential implications for forest governance. No matter how consultative and deliberative a participatory process may be, if such a process does not take into consideration the gender component, it is likely to generate negative consequences for the governance of forest resources.

Based on our empirical examination of participatory exclusion, we identified three major implications for forest governance in the Indian context. Firstly, the absence of women in GB meetings in KFPC affected the kind of rules framed for forest protection, especially regarding forest closure. During the initial years of forest protection, KFPC framed rules barring entry to the forests. The menfolk believed that banning entry to already degraded forests would not create any adverse effects. However, such a ban was found to have negative effects on women's movements inside the forest for fuelwood collection. Closure of the nearby forest by the FPC meant that women in the community had to look further afield for alternative sources of fuelwood, spending more hours in fuelwood collection and assuming the extra burden of carrying it from far-off places. Similar observations of the plight of women in JFM community forestry projects were also reported by Sarin (1998) and Agarwal (2001). Reflecting her concerns, one of the female KFPC members commented:

> The responsibility for collecting fuelwood for the household has always fallen on us. Since the menfolk started forest protection, the problem of getting fuelwood has become more acute. We know they are doing a good thing by protecting the forest, but we certainly need fuelwood for the household, and if we are not allowed to move about inside the forest, where are we going to get it?

By contrast, even though the EC of BFPC restricted entry into the forest in the initial years of protection, it did not put a blanket ban on fuelwood collection. Instead, it crafted a ground rule that fuelwood collection would be permitted only for the women members of the village (not outsiders) in head-load and not by any other means such as in bullock-cart or bicycle. Further, women were permitted only one head-load of fuelwood from the forest per day, consisting of only dried and fallen twigs or leaves. This comparison shows that, when women were involved in decision-making about forest governance, their needs, interests and priorities were included in the forest governance practices.

Secondly, by ignoring gender concerns, exclusionary participation in KFPC affected the practice of forest governance by generating asymmetries in information sharing between women and men. Since women were not part of decision-making in KFPC, they did not have access to information regarding forest protection practices for the village, despite being primary stakeholders due to their day-to-day dependence on the forest. Characteristic responses from KFPC women members were:

> Male members of the household don't discuss with us the decisions at the meetings of the FPC. How do we know what rules are being framed for forest protection?

> Sometimes we have to face embarrassing situations with forest guards, because we don't know the protection rules. Even if we ask our husbands about the decisions at the meetings, they typically respond, 'why do you want to know about external affairs?'.

> Maybe if we were part of the decision-making process, we would gain more information about forest protection activities.

The third way in which exclusionary participation affected forest governance in KFPC concerns efficiency in implementation of rules and monitoring of rule violation. Since women's needs, interests and priorities were not a part of decision-making, violation of rules by women were more likely in KFPC. Women's presence and participation in rule-making in BFPC reduced the possibility of rule violation as they themselves perceived the rules as fair and just. Further, the all-male patrol team of KFPC forest guards faced difficulties in catching and dealing effectively with female transgressors. By contrast, the BFPC forest patrol team, which included women, could more effectively deal with female rule breakers.

The undermining of gender concerns in participatory projects ultimately points to a policy failure in ensuring inclusion and empowerment. It is important to reiterate that one of the fundamental goals of participatory policies is to bring about social transformation by ensuring (gender) equity and providing an opportunity for the expression of voice in matters that affect everyday life. Exclusionary participatory models – like the one discussed here – denied such opportunities to women, and made gender inequality an accepted norm in participatory processes. Exclusion of women from decision-making does not create sufficient incentive for them to protect the resource, and therefore calls into question the sustainable use and governance of forests. Participatory forest governance projects, by excluding women, may result in the regeneration of forests and increased availability of forest produce for the community in the short run, but sustainability of such projects and long-term benefits remain difficult to achieve. It may, therefore, be argued that the long-term success of participatory forest policies not only depends on the outcomes (in terms of positive physical changes in resources) but also who participates in these exercises.

Explaining women's participation

The two cases discussed in this paper raise a pertinent question as to why women in certain context are more likely to be excluded than in others. The two cases exhibited varying degrees of inclusion of women in forest governance despite similar socio-economic, cultural and ecological contexts. The theoretical justifications of women's inclusion in resource governance certainly affirm the necessity of including women in forest governance for a number of reasons, including: their ideological and symbolic closeness with nature (Miles & Siva, 1993); their material dependence on forests due to the gendered division of labour (Agarwal, 2001); their special knowledge of species varieties and processes of natural regeneration (Agarwal, 1992); and the shared experience of exploitation by women's wombs and nature (Krishna, 2008). However, these arguments do not adequately explain the reason behind women's participation in some contexts and non-participation in others. Some empirical studies of women's participation in forest governance cite factors such as rules of entry, social norms of patriarchy and their personal characteristics as determinants (Agarwal, 1997; Locke, 1999; Narain, 1994). It is a widely acknowledged fact that patriarchal social norms restrict women's entry to the public sphere.

Although these norms apply in both the empirical cases, the BFPC case exhibited relatively higher level of women's inclusion in forest governance. One of the factors facilitating women's greater involvement in BFPC was their social network organisations in the village. The BFPC women used their social networks and sought help to gather courage to transgress the boundaries that separated the public from the private sphere. Women's collective mobilisation, which existed in the village for other social, economic and cultural issues, was strategically extended and used to ensure their presence in the public sphere as noted by one of the BFPC women:

> We support, encourage and help each other in matters of forest protection. Some of us do not have the confidence to stand up in public and speak in front of elderly men in the village. On those occasions, experienced and vocal women come to their help, and speak on their behalf.

Evidence of the collective mobilisation of women was not limited to forest protection activities in BFPC. A vibrant women's collective existed in this village for socio-cultural purposes. For instance, the women observed several cultural and religious festivals together, which built their capacity to exercise their agency in other situations. In a different context, scholars like Bardhan (2000) and Seabright (1997) highlight the self-reinforcing nature of cooperation, where cooperation in one sphere generates similar behaviour in other spheres. The social networks that have emerged in BFPC were replicated and reproduced ensuring women's greater involvement in forest protection activities.

The role of critical mass in collective mobilisation for resource management and in social movements is well documented in the literature (Marwell & Oliver, 1993). The findings from this study demonstrated how the presence or absence of a critical mass of women in institutions of forest governance impacts on women's inclusion in resource governance. Some of the emerging literature in the field also suggests that the number of women present in decision-making institutions plays a significant role in determining the overall participation of women (Agarwal, 2015). Women's greater inclusion in BFPC was due largely to the presence of a critical mass of women in both the EC and GB. The presence of women in relatively larger number boosted their confidence to stand up and speak for themselves in decision-making meetings. Overcoming the barriers of patriarchy and socio-economic, cultural and power-based gender inequalities was possible in the village due to the presence of increasing numbers of women in forest-governing institutions. In BFPC's general meetings, women sat together (separately from men) and talked among themselves. This created a comfort zone and paved the way for expressing their voices in decision-making. A woman from BFPC revealed that:

> I feel so good and comfortable when there are more women like me in the meetings. Men cannot ignore our concerns if there are more women to express their opinions. Further, being together and the companionship give me confidence. Even those women who cannot otherwise speak publicly also open up and express their concerns when in groups.

Patriarchal social norms, gender stereo-typical responsibilities of child bearing, rearing and household chores and the gendered division of labour as barriers to women's participation in the public sphere are also well documented in the literature. Our case study showed how rural women negotiated in patriarchal systems to contribute to forest protection activities. The BFPC women emphatically cited support and cooperation from husbands and mothers-in-law as one of the factors for their presence and participation in the forest governance institution. Rural women gained this support in patriarchal households not by explicitly contradicting their menfolk, but by negotiating and adopting bargaining strategies. One of the BFPC women stated:

> It is difficult for me to ignore my household chores to join the forest protection activities. In our society fetching fuelwood and water and preparing food for the household are women's jobs and we need to do them. Sometimes in order to go out to meetings or forest protection activities, I need to complete my household chores early. In this way, I gain support from my husband and mother-in-law.

The women's bargaining power depended upon their access to social networks. The presence of a critical mass of women increased their bargaining power. Some respondents stated that their husbands and mothers-in-law agreed to their participation only when other women in their group convinced them to do so. To quote a women member from BFPC:

> Initially my husband and mother-in-law were opposed to my attendance in forest protection meetings. But when other women members came to my house and assured them that they would be with me, my family members let me go. Now they do not object to my going out if I go with other women from the neighbourhood.

In other words, women's strategies for dealing with household responsibilities, support from family members, access to social networks and collective strength can help loosen restrictive social norms and facilitate women's participation in collective action. However, certain norms like the gendered division of domestic work remain rigid, but without necessarily creating barriers to women's participation.

Conclusion

The evidence of community involvement in participatory forestry projects suggests that even an otherwise successful participatory model bears the risk of excluding women. Our empirical observations substantiate the evidence from the available literature (Agarwal, 2010). When assessing participatory models of resource governance policies and practices, it is, therefore, not enough to examine the outcomes that participation produces for the community at large. Rather, who participates in participatory models emerges as an important indicator for evaluating the success of participatory policy. The formulation of participatory policies of resource governance, while being considered as necessary conditions for the creation of a space for people to have voice and to be regarded as stakeholders in governance, are not sufficient to avoid exclusionary practices.

Evidence from the case studies reported in this paper suggests that special efforts may be needed to address gender concerns in community-based resource governance if women are to be included in institutions of participatory forestry. Otherwise they are likely to remain excluded due to their multiple marginalities. Exclusionary participation, as reported in this paper and as observed elsewhere, would seem to have the potential to prevent the success of a well-designed participatory programme not only in community-based forest governance, but also in other spheres of governance and development, where active community participation is a policy objective. It is indicative from this study that achieving gender parity – a long-cherished policy objective – often requires greater access by women to social networks, the presence of a critical mass of women, the mobilisation of women collectives and increased bargaining power to negotiate with restrictive patriarchal social norms.

Acknowledgement

The author would like to thank Prof. Linda Hantrais and Prof. Ruth Kattumuri, London School of Economics & Political Science (LSE), London, UK; and the two anonymous referees for their comments and suggestions on the paper. Research support provided by Dr. Pratyusna Patnaik, National Institute of Rural Development & Panchayati Raj (NIRDPR), Hyderabad, India is duly acknowledged.

Disclosure statement

No potential conflict of interest was reported by the author.

Notes on contributor

Satyapriya Rout is an Assistant Professor in Sociology at University of Hyderabad, Hyderabad, India. He held a Sir Ratan Tata Fellowship at the Asia Research Centre, London School of Economics and Political Science, London, UK, during 2012–2013, where he carried out post-doctoral research on social inequality and access to water. His research areas span: environment and natural resource governance, environmental movement, decentralised governance and development, focusing on forest and water, and the question of community engagement with governance of natural resources.

References

Agarwal, B. (1992). Gender and environmental debate: Lessons from India. *Feminist Studies, 18*(1), 119–158.

Agarwal, B. (1997). Environmental action, gender equity and women's participation. *Development and Change, 28*(1), 1–44. doi:10.1111/1467-7660.00033

Agarwal, B. (2001). Participatory exclusion, community forestry and gender: An analysis and conceptual framework. *World Development, 29*(10), 1623–1648. doi:10.1016/S0305-750X(01)00066-3

Agarwal, B. (2010). *Gender and green governance: The political economy of women's presence within and beyond community forestry*. Delhi: Oxford University Press.

Agarwal, B. (2015). The power of numbers in gender dynamics: Illustrations from community forestry groups. *Journal of Peasant Studies, 42*(1), 1–20. doi:10.1080/03066150.2014.936007

Bardhan, P. (2000). Irrigation and cooperation: An empirical analysis of 48 irrigation communities in South India. *Economic Development and Cultural Change, 48*(4), 847–865. doi:10.1086/452480

Cuomo, C. (1998). *Feminism and ecological communities: An ethic of flourishing*. New York, NY: Routledge.

Krishna, S. (2008). *Genderscapes: Revisioning natural resource management*. New Delhi: Zuban Books.

Locke, C. (1999). Constructing a gender policy for joint forest management in India. *Development and Change, 30*(2), 265–285. doi:10.1111/1467-7660.00117

Marwell, P., & Oliver, P. (1993). *The critical mass in collective action: A micro-social theory*. Cambridge: Cambridge University Press.

McKean, M. A. (2000). Common property: What is it, what is it good for, and what makes it work? In C. C. Gibson, M. A. McKean, & E. Ostrom (Eds.), *People and forests: Communities, institutions and governance* (pp. 27–55). Cambridge: MIT Press.

Mellor, M. (2007). Ecofeminism: Linking gender and ecology. In J. Pretty, A. S. Ball, T. Benton, J. S. Guivant, D. R. Lee, D. Orr, … H. Ward (Eds.), *The SAGE handbook of environment and society* (pp. 66–77). London: Sage.

Miles, M., & Siva, V. (1993). *Ecofeminism*. London: Zed books.

Narain, U. (1994). *Women's involvement in joint forest management: Analysing the issues*. New Delhi: Ford Foundation.

Narayan, D. (1995). *The contributions of people's participation: Evidence from 121 rural water supply projects* (Environmentally sustainable development occasional paper series no. 1). Washington, DC: World Bank.

Paul, T. (1987). *Community participation in development projects: The World Bank experience* (World Bank discussion paper no. 6). Washington, DC: World Bank.

Phillips, A. (1995). *The politics of presence: The political representation of gender, ethnicity and race*. Oxford: Oxford University Press.

Poffenberger, M., & McGean, B. (Eds.). (1996). *Village voices, forest choices: JFM in India*. New Delhi: Oxford University Press.

Robinson, E. C. (1998). *Greening at the grassroots: Alternative forestry strategies in India*. New Delhi: Sage.

Sarin, M. (1998). *Who is gaining? Who is losing? Gender and equality concerns in joint forest management*. New Delhi: Society for Promotion of Wastelands Development (SPWD).

Seabright, P. (1997). Is cooperation habit-forming? In P. Dasgupta & K.-G. Mäler (Eds.), *The environment and emerging development issues* (pp. 283–306). Oxford: Clarendon Press.

Sundar, N., Mishra, A., & Peter, N. (1996). Defending the Dalki forest: Joint forest management in Lapanga. *Economic and Political Weekly*, *31*(45), 3021–3025.

White, S. (1996). Depoliticising development: The use and abuse of participation. *Development Practice*, *6*(1), 6–15. doi:10.1080/0961452961000157564

Community empowerment in changing environments: creating value through food security

Nicole Kenton and Sumita Singha

ABSTRACT

Methodologies that are locally relevant, empowering and replicable offer a way forward for improved policymaking and efficient resource governance in a globalised landscape of rapid change and scarce resources. Using the principles of equal participation and distribution, this article shows how the use of participatory methods can lead to greater community ownership and cohesion around shared concerns over access to healthy food and sustainable resource use in challenging urban environments. Drawing on contextualised examples from small-scale projects carried out by Charushila, an international environmental design charity, in Venezuela and Palestine, the article presents a co-design approach that puts people in touch with food growing and the reuse of resources to transform open spaces. An analysis of community-led co-production projects in these two contrasting urban environments shows how such processes can contribute to policymaking for longer-term sustainable development in the field of disaster relief and amid political upheaval in low- and middle-income countries.

Introduction

Local ownership is a crucial starting point for both individual and collective empowerment in projects seeking to bring about sustainable change. The small-scale case studies carried out in deprived urban settings in Venezuela and the occupied Palestinian territories presented in this article illustrate how communities can be empowered through food growing and positive environmental change in their neighbourhoods. The two projects have been selected from a larger number carried out by Charushila in India and the UK between 2010 and 2017 to exemplify the methodology and the challenges that can arise from food insecurities due to abiding political unrest. The case studies demonstrate that, while political resolution may take a long time, small-scale local initiatives and awareness-raising can be more quickly achieved and can offer longer-term solutions. The article explores how:

For more information on the projects described in this article, see http://charushila.org/completed-projects.html, https://www.facebook.com/charushila.org/, https://twitter.com/charushila_org

- people-initiated projects can be supported and facilitated;
- community gardening/food growing and enhanced food knowledge can contribute to greater food security and better nutrition;
- children and older people can be part of the process;
- community ownership and project continuity can be assured;
- environmental, cultural and social values relating to food growing can be enhanced through the involvement of art and artists.

The authors explain how an environmental design charity applied a community-led approach in low- and middle-income countries experiencing political challenges. They explore issues around assessing the impacts of small-scale interventions on the health and well-being of the communities targeted, taking gender and generation into consideration, and draw out lessons learned for evidence-based policy development. In conclusion, they consider whether such approaches can be replicated and how future policy planning processes at the local level might be adapted to achieve greater impact.

Conceptual and contextual frameworks

Food security is defined internationally as a condition in which

> all people, at all times, have physical, social and economic access to sufficient, safe and nutritious food that meets their dietary needs and food preferences for an active and healthy life. (FAO, 1996, Plan of Action, Point 1)

In 2016, the number of chronically undernourished people in the world was estimated to have increased to 815 million from 777 million in 2015 (FAO, IFAD, UNICEF, WFP, & WHO, 2017). Hunger statistics have historically focused on calorific intake, prompting nutritionists and researchers to advocate a multi-dimensional concept of food and nutrition security (Burchi, Fanzo, & Frison, 2011). Micronutrient deficiency imposes enormous costs on societies in terms of ill health, lives lost, reduced economic activity and poor quality of life (Shetty, 2011).

The nutritional aspect is integral to the concept of food security (CFS, 2009) and is achieved when the four main dimensions of food security – food availability, food access, food utilisation and stability – are fulfilled simultaneously (FAO, 2008). The double burden of malnutrition is characterised by the coexistence of under nutrition along with overweight, obesity, or diet-related non-communicable diseases (WHO, 2017).

Conflicts, increasingly exacerbated by climate change, are also on the rise, particularly in the middle East, and are severely affecting food security (FAO, IFAD, UNICEF, WFP, & WHO, 2017), bringing increased migration and a rural exodus. Increasing urbanisation means that more and more people are dependent on accessing food from within urban areas, which is also associated with dietary shifts towards more processed and pre-prepared foods, in part in response to long working hours and, for a proportion of the urban population, in combination with reduced physical activity (De Haen, Stamoulis, Shetty, & Pingali, 2003; Popkin, 2001). The rising consumption of preserved foods and drinks is not just about convenience and taste, but is also explained by economic, geographical and cultural reasons, as illustrated by the case studies in this article. In hot climates, such as in Venezuela, refrigeration is needed for fresh food and, if refrigerators are unaffordable, tinned or preserved foods can be convenient and economical alternatives. Tinned food can be standardised, transported without damage or deterioration, and sold in small units. Many forms of food processing are beneficial. However, ultra-processing, a type of process that has become increasingly dominant, at first in high-income countries, and now in middle-income countries, creates attractive, hyper-palatable, cheap, ready-to-consume food products that are characteristically energy-dense, fatty, sugary or salty and generally obesogenic (Monteiro, Moubarac, Cannon, Ng, and Popkin, 2013).

The interconnections between healthy eating, educational levels and economic status are also resulting in 'food deserts' where residents of low socioeconomic status are unable to access nutritious, affordable food in the vicinity of their home (Shrimpton & Rokx, 2012). Socioeconomic dynamics also come into play regarding food choices. According to an international study by Van Ansem, Schrijvers, Rodenburg, and van de Mheen (2014), children with highly educated mothers consume more pieces of fruit and more grams of vegetables per day and are more likely to eat breakfast than children with poorly educated mothers. So although poverty is undoubtedly one factor of undernutrition, nutritional education can be effective in making families aware of the importance of a healthy diet, and families can improve their nutritional health within the resources available to them (Khan, Rafique, Qureshi, & Badruddin, 2013). In addition, the packaging of processed foods and drinks contributes to environmental pollution, especially where refuse collection and recycling facilities are limited, as was the case in both the sites selected for the study.

Solutions to these global challenges lie not only in food production and access to knowledge about healthy eating but also in successful food policies that are informed by local knowledge. According to WHO/UN-Habitat (2016), problems such as obesity can be addressed by legislation that supports local food production, for example through informal markets, farmers' markets, community gardens and urban agriculture. Governments at all levels are being called upon to ensure that their policies are based on solid evidence from the ground, requiring community participation in decision-making processes and policy development to increase access to resources (McCullum, Desjardins, Kraak, Ladipo, & Costello, 2005).

Co-production, as conceived in the case study sites, is a process of active exchange and shared decision-making, and is an important long-term investment. It challenges the concept of a passive beneficiary by creating an expectation on the part of the service provider and the customer of equivalently active roles in the outcome of a project (Bettencourt, Ostrom, Brown, & Roundtree, 2002). For this reason, the projects presented in this article could be said to have three different roles: co-governance, co-management and co-production.

In an article on public policy, engagement and participation, Bovaird (2007) argues that projects designed within a long-term perspective should be based on the assumption that both citizen and the state have an equal contribution to make. Evidence of the value created in a community through a co-production approach, with its wide-reaching and interconnecting benefits, suggests that it should be at the heart of policies designed to build sustainable communities. This process is even more crucial in marginalised communities experiencing disrupted political processes, rapid change and scare resources, as in the urban areas in Venezuela and Palestine. In terms of improving food security, community-level initiatives can be shown to enhance food knowledge and increase interaction with food growing. These are, therefore, key policy considerations that can have a cascade of positive outcomes for individual and community well-being.

Food security in the case study countries

The case study sites – urban areas in Venezuela and Palestine – share common concerns such as food insecurity, fragmentation of grassroots participatory processes, difficult political landscapes and, most significantly, weak food distribution networks. Food insecurity is exacerbated by political instability (EIU, 2016). In 2015, the UN Food and Agriculture Organisation (FAO) praised Venezuela for having almost eradicated hunger, and for its 'food sovereignty experiment', involving agrarian reform and direct citizen participation in the food system (Schiavoni, 2015). Yet, at the same time, the country was experiencing food scarcity and associated riots. By 2017, the situation had worsened, with widening poverty, food and medicine shortages and alarming rates of child malnutrition (Caritas Internationalis, 2017).

Venezuela was the first country in South America to become a net importer of food using petro-dollars. For nearly a century, Venezuela's economy has been based on petroleum, which still accounts for 90% of its exports (The Economist, 2017). Since the 1930s, the twin focus on petroleum and urban development means that over 90% of the population today lives in cities. These conditions facilitated not only the development of an externally controlled framework of a food import and distribution complex but also the internal loss of knowledge of food growing, biodiversity and nutrition. In 2016, the humanitarian situation in Venezuela significantly deteriorated, as the impact of the drop in oil prices began to be more widely felt. In combination with income-restricted imports, the fall in oil prices has led to severe shortages of food and basic goods (ACAPS, 2017).

Researchers from the country's leading universities carried out a national living conditions survey (Encuesta Sobre Condiciones de Vida en Venezuela, ENCOVI) between 2014 and 2017 when the government's official household survey stopped being publicly available. The 2014 survey found that Venezuela was a country with two diametrically opposing forms of malnutrition: undernourishment and obesity, caused by a poor diet (International Crisis Group, 2015). It also noted corruption in the public food distribution network. The 2016 ENCOVI survey, based on interviews from a sample of 6413 households across the country, highlighted disparities in the quality and quantity of food, with access, availability and security being severely compromised (ENCOVI, 2016).

Palestine also had a disrupted food distribution system for a long time, owing to political restrictions on the movement of people and goods, notably in Gaza. The embargo and the closure of illegal tunnels in 2014 resulted in high prices for food and increasing levels of poverty. Food insecurity in Palestine remains severe, with more than one-quarter of the population (27% or 1.6 million people) categorised as food insecure (PCBS/FSS, 2016). Palestine's total dependence on Israel for access to essential goods and services, including fuel, means that food distribution has become a political issue. While Palestinian youth remain disengaged from governmental politics, they express their political needs through the NGOs because of the freedom that these platforms afford them, as was the case in the study: 27% of Palestinian youth connect with NGOs, a similar percentage as those affiliated to political parties (Ahmad, 2013).

Access to fresh food is a major challenge in Palestine for many reasons. Agricultural land appropriation by the Israeli occupation is a political cause for the loss of local food growing, leading to restricted imports of mainly processed foods via Israel. Israel's 'siege' or draconian closure of the Gaza Strip since 2006 has created conditions inimical to human life through resulting shortages of potable water, electricity and basic nutrition, and inaccessibility of essential medicines and medical care (HSRC, 2009). The occupation of lands deprives the Palestinian economy of 63% of the West Bank's agricultural resources, including the most fertile and best grazing land. Other physical barriers, such as walls, fences and no man's land (Category C land), have effectively decreased the available land, restricting it to a tight geographical space, while the expansion of Israeli settlements has further diminished the area available for agricultural activities (UNCTAD, 2015).

A study carried out in 2014 among mothers in the Gaza Strip showed obesity and overweight rates in urban areas, refugee camps and rural areas to be 57% (El Kishawi, Soo, Abed, & Muda, 2014, p. 3). Diabetes, which affects 20% of the population in Palestine, and consequent eye diseases, such as diabetic retinopathy, are on the increase, including among small children (St John of Jerusalem Eye Hospital Group, 2017), due to the consumption of sugary drinks by children and mothers during pregnancy. The consumption of fizzy drinks is high due to alcohol not being permitted for religious reasons, combined with the lure of soft drinks in a hot climate. Many other personal explanatory factors are also at play, such as the dynamics of the home environment, attitudes, cultural norms and lack of awareness of healthy eating.

The declining biodiversity of the land and lack of knowledge of food growing present similar challenges in both the Palestinian territories and in Venezuela. The absence of ecological knowledge and protection is a consequence of being deprived of opportunities to interact with nature and food growing. Both countries are also vulnerable to the impacts of climate change, which have affected agricultural production. Palestine is subject to extreme weather events and sea level rise (UNDP, 2011). The barrios (shanty towns) in Caracas are located on steep and unstable terrain, making the slopes particularly vulnerable to landslides associated with flooding and heavy rain (Wamsler, 2014).

The differences between the two countries are marked as well. Compared to Palestine, Venezuela has a more amenable climate, which means that, in theory, land can be cultivated with no restrictions on food production, making food security feasible (FAO, n.d.). Before the 1950s, agriculture, fishing and forestry were central to the Venezuelan economy, producing more than half the gross domestic product (GDP). As the petrochemical industry expanded rapidly in the 1970s and 1980s, agriculture continued to decline, and the Venezuelan government made several attempts at land reform involving the expropriation of private land from the so-called oligarchy. FAO (2014) reports show, however, that 84% of the 21.6 million hectares available are covered by permanent pastures, mostly unchanged over the past 20 years; leaving just over 3% of arable land. The current contribution of agriculture to Venezuela's GDP is 4% (Index Mundi, 2017).

Palestine is not characterised by the large economic and social gap experienced in Venezuela; even the refugees are supported through UNHCR programmes, and their living conditions are better than those in the Venezuelan barrios. In Palestine, the extreme dependence on the political process for daily necessities sets off political debates, but they are voiced through the NGOs that work there, rather than through media (Ahmad, 2013).

Case study methodology: creating value

The methodology adopted in the case studies is based on 'value creation', the positive transformation of reality, a theory developed by the Japanese philosopher, Tsunesaburo Makiguchi, and achieved through the three aspects of 'beauty, benefit, and good'

(Bethel, 1989). The 'good' is represented by the cascade effect of the positive aspects of the projects on the community, while the use of art construes the 'beauty' aspect, and 'benefit' is the empowerment of the individual entwined with the empowerment of the community. The creation of value through the projects is not an abstract concept or separated from the community or the individual; it is intrinsic to the overall goal of food security and healthy living. Individual tactics or actions for each project, therefore, comprise many different strands, including three essential practices: the involvement of young and older people, as well as artists; the use of waste and found products and the growing of food. Other actions, which are more fluid and depend upon the specific nature of the project, are exemplified in the two case studies.

In both study sites, an initial quantitative and qualitative survey of the community's needs was carried out using questionnaires, semi-structured interviews and participatory workshops, in collaboration with local partners, including municipalities and universities. Working with local partners and combining methodologies provided additional information about the local demographics and a greater understanding of the health of the community.

Values were assessed in interviews and 'before and after' surveys designed to identify the positive social, cultural, and environmental effects of the projects. Social value is concerned with how places and buildings encourage people to interact in ways that lead to trust, mutual understanding, shared values and supportive behaviour (Commission for Architecture and the Built Environment, 2006). A key indicator of social value is the proportion of people who feel there is a good community spirit where they live. Such social values are difficult to measure using quantitative methods alone, since the rich details of quality may be missed.

A range of stakeholders participated in the initial feasibility assessments and engagement, including residents and local NGOs. The workshops provided an opportunity for the local community to discuss their ideas openly and in more detail, and to shape their vision for the spaces collectively. Participants in both areas cited food growing as a top priority, along with a safe space to play and relax, amid concerns about anti-social behaviour or gang-related violence. Consultations and feedback sessions continued throughout the project cycle with further workshops and presentations, and the final design was adjusted accordingly. In all cases, one or two key local community members took charge of the running of the initiatives during the projects and after they had ended. The project's value creation methodology came into play through drawing, painting, pottery and seed-growing workshops, and community gardening.

The approach was adapted according to the context, as in the other countries where the charity has worked, such as the UK and India. Table 1 demonstrates the contextual similarities and differences between the two case studies and countries. Both projects, for example, involved many non-governmental or charitable organisations, but the incidence of NGOs is much higher in Palestine than in Venezuela, and numbers of those who participated were also higher in Palestine. Economic and gender disparities in Venezuela were clearly reflected in those who engaged in the project.

The Venezuelan project

In Venezuela, the project was undertaken with a community group based in the barrio of San Agustín, Caracas. The aim was to create a community garden from an area of

Table 1. Comparison of the case study projects.

Description	Venezuela (2010)	Palestine (2013)
Religion (predominant)	Catholic Christians	Islam
People engaged	Total: 43 17 adults 26 children	Total: approximately 150 40 children + 30 teenagers 80 adults
Stakeholder partner organisations	Universidad Central Venezuela (UCV, public sector) Metrocable (Public sector) La Ceiba school (Public sector) Military and police (Government) Odebrecht Organisation (Brazilian construction company)	Revivers of the land of Palestine (NGO) Birzeit University (Public sector Al Bireh Local Authority (Public sector) Local shops (private) Al-Amara refugee camp and school (UNHCR supported) AMIDeast (NGO) RIWAQ (NGO) British Council (British cultural organisation) Palestine Green Building Council (NGO) Dalia Association (NGO)
Type of project	Kitchen garden	Kitchen garden, seating and environmental improvements
Costs (approximate) (no labour costs)	£430 (most private donations)	£150 (most public donations or in kind)
Size of site	One site of 2 acres reclaimed space in barrios, out of the planned urban area	Two sites: 0.5 acres and 0.2 acres (public land inside city)
Impact/further community projects without support	Three projects	Over 100 projects

wasteland. The community called the garden 'El Shanti' (from the Sanskrit word meaning 'peace'), the overall objective being long-term community unity and peace. Discussions with community members prioritised the following aspects that were revealed during the year-long consultation period: growing locally produced food, organic and low-cost production, ownership and maintenance by the community, reuse of waste and rubble, stabilisation of the slope and soil, a space for relaxation and recreation, engaging young people in the design of the space, raising community self-esteem (often associated with body image among young women), and community unity.

The barrio of San Agustín is located across a hill around the south of Caracas. The residents do not have convenient access to fresh foods due to the long distance from city shops. Nor do they have the physical strength required to haul groceries up or down the equivalent of several flights of stairs, necessitating the use of more 'long-lasting' produce, such as chemically preserved, dried and canned foods (Sinha, 2012). Small-scale food production in the barrios thus offered a pathway towards eating fresh foods and community self-sufficiency. During the year-long consultation, several exchange visits took place between the community garden and an organic farm in the city centre of Caracas. Techniques learnt from the farm, which itself was inspired by inner city food growing in neighbouring Cuba, included using *organopónicos*, an organic substrate obtained from crop residues, household waste and animal manure, which can be applied on building sites, vacant lots and roadsides, and arranged in terraces on sloping land (FAO, 2014).

The site chosen for the community garden was an area where homes had been demolished to build the city's first cable car system (Metrocable) from the centre of Caracas up to the barrio, which was previously only accessible by foot. Building material had been dumped on the site, creating another potential hazard in a landslip area. Reuse of these materials, adopting a gabion construction, helped to stabilise the soil and slopes. Due to the location of the garden, removal of the demolished materials was difficult. It could not be taken on the Metrocable, so reuse was also a necessity. The community conceived the idea of wrapping the rubble in chicken wire mesh (the 'rubble tortilla' as it came to be called) and creating planters from the units.

The focus on young people was crucial for the future well-being of the community. As a resident in Caracas put it:

> I want to see children grow up without violence and drugs and [to experience] educational training and environmental awareness. Where these gardens will flourish in communities of indescribable beauty, we will too.

In the barrios, few areas exist for recreation for children, apart from the streets. Teaching young people to grow food and eat healthily was therefore a major objective. The local partner organised a children's drawing workshop, which attracted 26 children and produced colourful and inspired artwork, some using stones and leaves from the site. The ideas generated proved to be a catalyst for the garden design, which became an iterative, community-led process. In an engagement workshop, one girl voiced her aims for doing such work:

> I want to bring this knowledge to my high school, so my classmates can spread this information to their communities and regenerate our planet, including people, animals, plants, objects and us. We are the only living things with the intellectual capacity to do so.

In follow-up meetings, the community expressed not only a sense of physical ownership due to the location of the garden, but also an ongoing hands-on relationship. Owning and using the space in a positive way provided a focal point for unity in the barrio, which had become the site of drug-related violence. As one middle-aged female participant stated: 'I include myself in the project to learn more and also to give my knowledge. I am prepared to learn from you'. A safe space to meet, grow food and play with their children was very important to the women. They could make concerted efforts to tackle unhealthy eating in a supportive environment. Gardening also provided an inclusive physical activity. Older people could enhance their mobility and relax. Raised beds for planting were added for their ease. Accessible green spaces, coupled with involvement in local community activities, have been linked to longevity among older people in densely populated urban areas (Takano, Nakamura, & Watanabe, 2002).

The Palestinian project

The Palestinian project was organised in the urban areas of Al Bireh, a city joined with Ramallah, on the West Bank. The project team was approached by local organisations – Ommar El Ard (Revivers of the Land of Palestine) and AMIDeast (an NGO supported by USAID) – who were working in the area. They had similar environmental goals and objectives, such as promoting local produce, self-sufficiency and an ecological response. Ommar El Ard was involved in delivering environmental education towards a clean and healthy Palestine, working with schools and arranging neighbourhood clean-ups with residents, volunteers and local municipalities. Another organisation, the Dalia Foundation, was advocating for the rights of Palestinians to control their own resources and sustain their own development for future generations, providing grants to communities and encouraging giving through philanthropy and volunteering.

The local organisations, volunteers from the municipality and an environmental student group from Birzeit University all expressed their wish to create a green and sustainable Palestine. They were concerned about access to healthy food and the well-being of community members. They were committed to cleaning up the area and raising awareness of recycling by encouraging the community to get involved in clearing spaces and finding solutions for waste management. One volunteer from Birzeit University said that the best way to make Palestine green and sustainable was by 'raising awareness in the community and getting them involved, not leaving it for the NGOs'. Another independent volunteer suggested the solution would be to 'encourage local produce and production, creating a circular economy and reusing waste'.

Initially, the charity was asked to work on five sites. The two sites chosen were in built-up zones used as rubbish dumps and unsafe. Together with children from a refugee camp and local community members, 32 bags of rubble and waste were cleared. The waste materials were reused where possible. An art workshop was organised where teenagers and young children drew pictures and verbalised their concepts of a sustainable Palestine, which became encapsulated in the design of the spaces. One teenager asserted: 'I am not happy with the pollution generated with the cars', and proceeded to draw a picture of tree-lined streets as a vision of sustainable Palestine.

Community at work in the gardens created in Al Bireh, Palestine (credit: Nadia Habash).

Since access to food and general mobility were restricted in the area due to Israeli occupation and restrictions, putting the land to productive use was an expressed wish from the community. In total, around 150 people took part in the process of building and planting in the open public spaces. Rubble, stones, disused tyres, plastic bottles, and other waste materials were used to create a stylised garden. Mindful of the acute scarcity of water in the area, and of restrictions on the use of rainwater harvesting equipment by the Israelis, the children grew drought-tolerant herbs, and rescued plants were replanted. Other art workshops, led by a local artist, were organised subsequently to draw murals. On the retaining walls in one of the sites, children and teenagers painted Palestinian flags and slogans. One of the youngest helpers was four-year-old, Ahmet, who came every day after school to help.

As the project became embedded in the community, renewed emphasis was given to healthy living through eating local and seasonal foods, and to recycling initiatives in urban areas. The local community took charge of the completed spaces, and the municipality agreed to support such projects in the future. The authorities declared the area to be of special importance, and it was featured in local newspapers.

Challenges to project management

In both projects, challenges arose around issues of regular and continued community participation. In Venezuela, some community members, especially the men, did not turn up for meetings or for work on the site. The women who worked for the Metrocable company remained committed to the project, even when they were not paid their salaries due to an industrial dispute. The weather was adverse: it rained heavily throughout the project cycle, with thunderstorms and landslides (about 2000 houses collapsed in Caracas, 600 in San Agustín alone with three fatalities). More support was then forthcoming from external partners, including a construction company and researchers from the Central University of Venezuela (UCV), and the work was planned in consultation with the community.

A follow-up community meeting facilitated by UCV gathered feedback, and the lack of commitment to the project by some community members was discussed. The low participation of men and young people and the ongoing violence in the community were recognised as challenges. However, those involved in the new projects observed a 30% reduction in crime since the start of the community engagement, according to 2010 figures from the architects Urban-Think Tank (U-TT) who were engaged in the Metrocable project.

In Palestine, a year was needed to plan the intervention due largely to the difficult political climate. Other challenges came both from building with very few resources and from needing to maintain the enthusiasm of the people involved. The lack of water and restrictions on rainwater collection initially threatened to impede the continuity of the project and its delivery. International companies provided bottled water in areas where drinking water was scarce, but it was then necessary to deal with the plastic bottles that were littering the environment. They were used creatively in the planting of the garden, together with abandoned tyres. Ultimately, the larger of the projects became unusable because the neighbouring Israeli colony increased its size, thereby reducing the available Category C land. Altercations between the Palestinians and Israeli soldiers meant that local NGOs had to go elsewhere to work.

Learning from the evidence base

Despite the challenges, follow-up with members of the community in the two sites, through feedback sessions, questionnaires and individual discussions, provided evidence of the many positive benefits emanating from the projects, at both individual and community levels. Participants mentioned increased physical activity, community cohesion and social interaction, and the consumption of more fresh fruit and vegetables. Local users felt motivated to maintain and improve the transformed spaces, as well as tackle wider issues, such as healthy living and waste management, to extend the benefits to the whole community.

The use of recycled materials and the interest in various rare medicinal plants that were found on the site by the community in Caracas helped to consolidate interest in extending an eco-tourism project. In both countries, recycling is rare, so the projects provided an example for others to follow. Helping to develop the area as collective owners was found to enhance the self-esteem of all generations of community members, empowering them to play a more proactive role in improving their well-being. The inclusion of children and young people in the projects, through art workshops and gardening, proved to be an empowering factor for them. They learnt about seeds and planting, and their input helped to shape the final garden designs, thereby nurturing potential community activists.

The environmental, cultural and social values intrinsic to the different communities were enhanced through work with art and artists. In both sites, local artists were brought in to hold workshops and produce final designs. In Venezuela, the art was innovative and practical. In Palestine, it was political; the community painted images, slogans and flags in vibrant colours. Art-based workshops across generations proved to be a powerful unifying force. The outcomes of the projects in these two diverse settings and cultures demonstrated that food growing, community engagement and environmental activism can be concordant and motivate active citizenship. In the short term, benefits were apparent in enhanced individual and community well-being, as expressed within the community.

The experience gained from the projects confirmed the importance of a long-term consultation process, and of using a phased approach to include space and time for the community to adapt and reassess the design, and for community 'champions' to build on the momentum generated by the projects. Through the process of co-designing the gardens, community members became equal partners in the projects. The early identification of local partners and community champions was crucial for continuity and continued ownership by the community. Bringing in donors and local organisations at key events to meet the community made future communication easier, and secured additional funding. The engagement of local NGOs and charities with relevant and complementary skills and experience at the outset provided ongoing support in the form of training and resources.

In a broad sense, the projects demonstrated the value of co-production in reinforcing healthy living and community engagement. Growing brings people together on many different planes: health, local produce, exercise and a sense of belonging (Public Health England, 2014). With climate change potentially leading to increased migration, wars and conflicts, simple and small acts of growing can help to build resilience among displaced communities, as shown in Palestinian, despite years of political oppression, or in Venezuela, where large numbers of people are migrating to Brazil and other neighbouring countries to feed themselves. Growing food proved to be not only a way of building resilient communities but also a compelling reason to pursue person- and community-centred approaches to health and well-being. It can also improve employment prospects and soft skills, as exemplified by the Venezuelan case.

Conclusion

This article has explored how small-scale people-initiated projects can be supported and facilitated locally; how community gardening/food growing and enhanced food knowledge can contribute to a greater degree of food security and improved nutritional practices at the local level; how children and older people can be part of the process; and how environmental, cultural and social values can be enhanced through working with art and artists.

In both the project areas, top-down bureaucratic and political interventions (or even consultation) had not hitherto resulted in strong community mobilisation. The two cases demonstrate that, when pursuing policies for the sustainable development of communities, a community-led and participatory approach may be more effective for implementation than institutional, managerial and technological methodologies. The cases provide evidence in support of the observation that, to be successful, food policies need to address the complex challenges of rising inequality, continued poverty, globalisation, climate change, food price volatility, natural resource degradation, demographic and dietary transitions (Pinstrup-Andersen & Watson, 2011).

The challenge in countries such as Venezuela and Palestine, which are both beset with political upheavals, is to find ways of effectively filtering proposals and reforms down to the grassroots, trying and testing them in situ, and then formalising them in policies. The testing process is often missing before formalisation. Although governments increasingly recognise the need for food security and acknowledge that the necessary evidence exists, the study cases show how important it is to harness local knowledge and to devote attention to the links and steps from the ground up.

A major challenge for the project team was to assess whether the Venezuelan and Palestinian projects had ensured continuity and community ownership. Ideally, ongoing monitoring should have been part of the remit of local partners. As the political situation in Venezuela deteriorated, and since the political turmoil in Palestine did not abate, it was not possible to return to the sites to carry out any *post hoc* evaluations, or to check progress against the initial indicators to determine whether any positive changes had been sustained. Feedback during the projects did show, however, that the new garden in St Agustín had become a valuable community resource. Small-scale food production was contributing to a healthier diet for those living in the barrio by providing fresh produce such as tomatoes and salads. The community planned to have an exhibition space about food growing, and the Metrocable project had an elected representative in San Agustín responsible for 'health, cleanliness and culture' (Icon, 2010).

In Palestine, various local community groups went on to improve other urban spaces and wastelands and transform them into gardens and playgrounds. One organisation received funding from international companies to beautify school playgrounds and promote active waste management, including a paper recycling project. Another campaigned for local and organic produce and income generation from selling donated goods. Yet another continued to work with teenagers in the city and refugee camps.

According to one of the NGOs: 'The initial projects ... set off a chain of other schemes, only temporarily halted by the 2014 Gaza war.' Despite the temporary inactivity due to the war, over 100 related activities were carried out.

A comment in the *Washington Post* suggests that the message conveyed by the Venezuelan project about the value of local initiatives for food security and community well-being may have had a wider impact: 'Some Venezuelan city dwellers are trying to grow their own produce to offset the country's severe shortages following socialist President Nicolás Maduro's calls for "food sovereignty"' (Barbarani, 2016).

As demonstrated in this article, strengthening the capacity of people to create value can bring about positive change in a community, using modest financial resources, and working with local partners to ensure sustainability through ongoing monitoring and evaluation. Makiguchi stressed the importance of human agency in enhancing the personal and collective lives of people when he claimed that: 'Individual well-being entails cooperative and contributive existence within society' (Makiguchi, 1983–1988, vol. 5, p. 129). Given that many wider aspects of food security and nutrition also need to be addressed, it can be said that empowering communities to take local food security into their own hands and providing strong evidence from the ground is a way of ensuring an equitable distribution of resources and building the foundation for a sustainable future, even in unfavourable and unsupportive political climates.

Acknowledgements

The authors are grateful to all project partners, volunteers and local community members in Venezuela and Palestine. In Venezuela, we would particularly like to thank the Central University of Venezuela (UCV), the artist Natalya Critchley and the local community organisations. In Palestine, we would like to thank the non-profit organisations AMIDeast and Ommar El Ard (Revivers of the Land of Palestine); the Dalia Association; children from the Al-Amari refugee camp; RIWAQ, the architectural conservation centre; the British Council and the environmental student group from Birzeit University.

Disclosure statement

No potential conflict of interest was reported by the author.

Notes on contributors

Nicole Kenton is a community development practitioner and an editor specialising in international development, with a particular focus on inclusive and participatory approaches. She was previously co-editor of the journal *Participatory Learning and Action*, published by the International Institute for Environment and Development (IIED). She became a director and trustee of Charushila in 2010.

Sumita Singha is an architect specialising in the design of healthy places. She is the author of several books including *Architecture for Rapid Change and Scarce Resources* (Sinha, 2012) and has taught architecture for over 20 years. She is also currently a non-executive director of Moorfields NHS Foundation Trust and a member of the Sustainable Development Unit at Public Health England. She founded Charushila, a registered charity working in environmental projects for communities in 2010.

ORCID

Nicole Kenton http://orcid.org/0000-0003-0798-6602
Sumita Singha http://orcid.org/0000-0001-7505-4132

References

ACAPS. (2017). Humanitarian needs in Venezuela. Briefing note. Retrieved from https://reliefweb.int/report/venezuela-bolivarian-republic/acaps-briefing-note-venezuela-humanitarian-needs-economic.

Ahmad, A. (2013). Palestinian youth: Between patriarchy and politics, *Atkin Paper Series*. ICSR. Retrieved from http://icsr.info/wp-content/uploads/2013/09/Atkin-Paper-Series_Ayed-Ahmad.pdf

Barbarani, S. (2016, November 1). Venezuela is telling hungry city dwellers to grow their own food. *Washington Post*. Retrieved from https://www.washingtonpost.com/news/worldviews/wp/2016/11/01/venezuela-is-telling-hungry-city-dwellers-to-grow-their-own-food/?utm_term=.bbfd0b8f7b5c

Bethel, D. M. (1989). *Education for creative living: Ideas and proposals of Tsunesaburo Makiguchi*. Ames: Iowa State University Press.

Bettencourt, L. A., Ostrom, A. L., Brown, S. W., & Roundtree, R. I. (2002). Client co-production in knowledge-intensive business services. *California Management Review*, *44*(4), 100–128. doi:org/10.2307/41166145

Bovaird, T. (2007). Beyond engagement and participation: User and community coproduction of public services. *Public Administration Review*, *67*(5), 846–860. doi:10.1111/j.1540-6210.2007.00773.x

Burchi, F., Fanzo, J., & Frison, E. (2011). The role of food and nutrition system approaches in tackling hidden hunger. *International Journal of Environmental Research and Public Health*, *8*(2), 358–373. doi:10.3390/ijerph8020358

Caritas Internationalis. (2017). *Caritas on Venezuela* (blog 16 August). Retrieved from http://www.caritas.org/2017/08/caritas-on-venezuela/

Commission for Architecture and the Built Environment (CBE). (2006). *The value handbook. Getting the most from your buildings and spaces*. London: Cabe.

Committee on World Food Security (CFS). (2009). *CFS reform document*. Rome: FAO. Retrieved from http://www.fao.org/fileadmin/templates/cfs/Docs0910/ReformDoc/CFS_2009_2_Rev_2_E_K7197.pdf

De Haen, H., Stamoulis, K., Shetty, P., & Pingali, P. (2003). The world food economy in the twenty-first century: Challenges for international co-operation. *Development Policy Review*, *21*(5–6), 683–696. doi:10.1111/j.1467-8659.2003.00232.x

The Economist. (2017, April 6). *How Chávez and Maduro have impoverished Venezuela*. Retrieved from https://www.economist.com/news/finance-and-economics/21720289-over-past-year-74-venezuelans-lost-average-87kg-weight-how?zid=305&ah=417bd5664dc76da5d98af4f7a640fd8a

Economist Intelligence Unit (EIU). (2016). *Global food security index 2016: An annual measure of the state of global food security*. Retrieved from http://foodsecurityindex.eiu.com/

El Kishawi, R. R., Soo, K. L., Abed, Y. A., & Muda, W. (2014). Obesity and overweight: Prevalence and associated socio demographic factors among mothers in three different areas in the Gaza Strip-Palestine: A cross-sectional study. *BMC Obesity*, *1*(7). doi:10.1186/2052-9538-1-7

ENCOVI. (2016). *Encuesta sobre Condiciones de Vida de Venezuela 2016: Alimentacion*. Retrieved from http://www.fundacionbengoa.org/noticias/2017/encovi-2016.asp

FAO. (1996). *Declaration of world food security*. Rome: FAO. Retrieved from http://www.fao.org/docrep/003//w3613e/w3613e00.htm

FAO. (2008). *An introduction to the basic concepts of food security*. Rome: FAO.

FAO. (2014). *Growing greener cities in Latin America and the Caribbean: An FAO report on urban and peri-urban agriculture in the region*. Rome: FAO.

FAO. (n.d.). *Venezuela. FAOSTAT*. Rome: FAO. Retrieved from http://www.fao.org/faostat/en/#country/236

FAO, IFAD, UNICEF, WFP, & WHO. (2017). *The state of food security and nutrition in the world 2017: Building resilience for peace and food security*. Rome: FAO. Retrieved from http://www.fao.org/state-of-food-security-nutrition/en/

HSRC. (2009, May). *Occupation, colonialism, apartheid? A re-assessment of Israel's practices in the occupied Palestinian territories under international law*. Cape Town, South Africa.

Icon. (2010, July 8). *Urban think tank's Caracas cable-car*. Retrieved from https://www.iconeye.com/architecture/features/item/4430-urban-think-tank-s-caracas-cable-car

Index Mundi. (2017). Venezuela economy profile 2017. Retrieved from http://www.indexmundi.com/venezuela/economy_profile.html.

International Crisis Group. (2015, July 30). *Venezuela: Unnatural disaster. Crisis Group Latin America Briefing.* N° 33. Retrieved from http://www.eluniversal.com/nacional-y-politica/150502/venezuelans-are-both-obese-and-malnourished

Khan, A. Z., Rafique, G., Qureshi, H., & Badruddin, S. H. (2013). *A nutrition education intervention to combat undernutrition: Experience from a developing country, ISRN Nutrition.* Article ID 210287. doi:10.5402/2013/210287

Makiguchi, T. (1983–1988). *Makiguchi Tsunesaburo zenshu [Collected works of Tsunesaburo Makiguchi]* (Vols. 1–10). Tokyo: Daisan Bunmeisha.

McCullum, C., Desjardins, E., Kraak, V. I., Ladipo, P., & Costello, H. (2005). Evidence-based strategies to build community food security. *Journal of the American Dietetic Association, 105*(2), 278–283. doi:10.1016/j.jada.2004.12.015

Monteiro, C. A., Moubarac, J. C., Cannon, G., Ng, S. W., & Popkin, B. (2013). Ultra-processed products are becoming dominant in the global food system. *Obesity Reviews, 14*(Suppl 2), 21–28. doi:10.1111/obr.12107

Palestinian Central Bureau of Statistics (PCBS), & the Food Security Sector (FSS). (2016, May). *Socio-economic & food security survey* 2014. State of Palestine. Retrieved from http://fscluster.org/sites/default/files/documents/sefsec2014_report_all_web.pdf

Pinstrup-Andersen, P., & WatsonII, D. D. (2011). *Food policy for developing countries: The role of government in global, national, and local food systems.* New York: Cornell University Press.

Popkin, B. M. (2001). The nutrition transition and obesity in the developing world. *Journal of Nutrition, 131*(3), 871S–873S. Retrieved from http://jn.nutrition.org/content/131/3/871S.full

Public Health England. (2014). *Sustainable, resilient, healthy people & places: A sustainable development strategy for the NHS, public health and social care system.* Cambridge: Sustainable Development Unit, Public Health England.

Schiavoni, C. (2015). The Venezuelan food sovereignty experiment. *Solutions, 5*(6), 46–53. Retrieved from http://www.thesolutionsjournal.com/node/237281

Shetty, P. (2011). Addressing micronutrient deficiency to achieve nutrition security. In B. Thompson & L Amoroso (Eds.), *Combating micronutrient deficiencies: Food-based approaches* (pp. 28–40). Rome, Oxford: FAO, CABI. Retrieved from http://www.fao.org/docrep/013/am027e/am027e.pdf.

Shrimpton, R., & Rokx, C. (2012). *The double burden of malnutrition: A review of global evidence* (Health, Nutrition, and Population (HNP) Discussion Paper). Washington, DC: Human Development Network, World Bank.

Sinha, S. (2012). *Architecture for rapid change and scarce resources.* London: Routledge.

St John of Jerusalem Eye Hospital Group. (2017). *Jerusalem hospital.* https://www.stjohneyehospital.org/our-impact/jerusalem-hospital

Takano, T., Nakamura, K., & Watanabe, M. (2002). Urban residential environments and senior citizens' longevity in megacity areas: The importance of walkable green spaces. *Journal of Epidemiology and Community Health, 56*(12), 913–918. doi:10.1136/jech.56.12.913

UNCTAD. (2015). *The besieged Palestinian agricultural sector.* Geneva: United Nations.

UNDP. (2011). *Climate change adaptation strategy and programme of action for the Palestinian authority.* Jerusalem: United Nations Development Programme.

Van Ansem, W. J., Schrijvers, C. T., Rodenburg, G., & van de Mheen, D. (2014). Maternal educational level and children's healthy eating behaviour: Role of the home food environment (cross-sectional results from the INPACT study). *The International Journal of Behavioral Nutrition and Physical Activity, 11*, 665. doi:10.1186/s12966-014-0113-0

Wamsler, C. (2014). *Cities, disaster risk and adaptation.* Routledge. doi:10.4324/9780203486771

WHO. (2017). The double burden of malnutrition: Policy brief. Retrieved from http://www.who.int/nutrition/publications/doubleburdenmalnutrition-policybrief/en.

WHO/UN-Habitat. (2016). *Global report on urban health: Equitable, healthier cities for sustainable development.* Retrieved from http://www.who.int/kobe_centre/measuring/urban-global-report/en/

Environmental impact assessment: evidence-based policymaking in Brazil

Kathryn Hochstetler

ABSTRACT

Environmental impact assessment (EIA) procedures aim prospectively to collect evidence about the environmental impacts of economic projects and to avoid or compensate for the costs incurred. This article asks whether such procedures have been effective in Latin America after many regional countries returned to some version of the developmental state after 2000. It does so by surveying the procedural effectiveness of Latin American regulations comparatively before turning to a deeper study of the Brazilian case. In Brazil, which has some of the strongest EIA procedures in the region, it finds that stakeholders make very different assessments of its effectiveness, not least because they define the standard differently. Economic actors in and out of the state criticise Brazilian EIA as ineffective from a transactive standpoint, which questions the time and cost associated with environmental licencing. Environmental and community activists see EIA as ineffective in achieving the substantive sustainability ends they value. Neither appreciates the procedural improvements offered by licencing professionals. The article concludes that EIA invites a broader set of stakeholders than did classic developmental states, but cannot on its own adjudicate among the resulting multiple visions of how to carry out development strategies.

Environmental impact assessment (EIA) has become one of the most common environmental regulations in the world (Morgan, 2012, p. 6). The heart of EIA is a process that accumulates and evaluates information about the likely impacts of proposed economic projects on the physical environments and local communities that host them (Glassen, Therivel, & Chadwick, 2012). The EIA process requires that someone – usually the firm carrying out a project – prospectively consider how a project is likely to change baseline environmental conditions in its area of impact. Much of this information is provided by natural scientists and engineers, but a full analysis of impacts requires social scientific analysis to assess social and economic changes as well. This evidence-based process operates with a view to preventing or ameliorating the negative impacts of the project and proposing compensation for the unavoidable harms. Those assessments are then reviewed by a separate actor, most often an environmental agency, that may grant or deny an environmental licence for the project. EIA processes also commonly include some form of consultation with the local host community (Hochstetler & Tranjan, 2016, p. 500).

The article begins with a discussion of how EIA might intersect with recent economic development strategies in Latin America. In many countries of the region, EIA processes are the main opportunity for public discussion of the advantages and disadvantages of specific development projects. As a result, EIA is an important locus for viewing how socio-environmental evidence becomes an input into decision-making and how policy-makers respond to such evidence. I argue that EIA is potentially compatible with developmentalist strategies, which aim to create the underpinnings for long-term national development, as sustainability is crucial for that goal. Nevertheless, EIA introduces a set of considerations and actors that have historically been left out of economic policy-making. Conflict is a common result, not least about whether EIA is effective and what effective EIA requires.

This general introduction is followed by a comparative analysis of current EIA processes in Latin America that provides insights into how regional governments collect environmental evidence for their decision-making. The comparison is based on a number of textual indicators that allow asking which of the regional EIA systems is the most procedurally effective, that is how closely legal texts follow expected 'standards and practices' (Chanchitpricha & Bond, 2013, p. 66). A fuller consideration of effectiveness requires more detailed empirical investigation, as provided by a case study of Brazilian EIA. The article's overarching conclusion is that even EIA processes that are comparatively strong technically remain enmeshed in political contestation over national development strategies in Brazil and Latin America, perhaps especially when states are active participants in planning and carrying out development projects (see also Barandiaran, 2015; Jaskoski, 2014). These conflicts reflect the different perspectives stakeholders bring to the question of what constitutes an effective EIA (Chanchitpricha & Bond, 2013).

Twenty-first century developmental states: what role for environmental evidence?

After a generation of neo-liberal rollbacks, many Latin American states have returned to playing a more central role in their national political economies after 2000, akin to an earlier generation of developmental states (Johnson, 1982). They built hydroelectric dams, roads and other infrastructure projects on a grand scale. They encouraged more extraction of oil, minerals and other natural resources in light of the highest commodity prices in decades. They offered public finance for private projects and contracted private firms to carry out some of their ambitious plans (Montero, 2014; Weyland, Madrid, & Hunter, 2010). Many countries in the region had also adopted new environmental regulations in the 1990s, including EIA regulations requiring them to consider systematically the effects of projects on the physical environments and local communities where they are to be located (Hochstetler, 2012). The post-2000 return to developmentalism thus marked the first intersection of EIA and state-led development strategies in the region. As such, it offers an excellent opportunity to view the nexus between environmental evidence and economic policy-making in an important, middle-income region.

Developmental states, in their ideal typical form, are 'defined by state intervention that is technocratically oriented towards the long-term development needs of the whole economy, even determining which industries should rise and fall' (Hochstetler & Tranjan, 2016, p. 498). The theoretical construct of the developmental state, which grew out of Johnson's (1982) path-breaking study of Japan, upheld the ideal of a state bureaucracy that was able to see beyond particularist interests to the larger economic interests of society as a whole (Wade, 1990, p. 29). In order to maintain this strict focus on the development goal, elite bureaucrats typically made economic decisions in a political system that was either fully authoritarian or where political actors and civil society were shut out of economic decision-making (Johnson, 1982, pp. 17–22). The classic developmental states considered little information from actors like environmental scientists, social scientists or local communities (Hochstetler & Tranjan, 2016, p. 500).

Classic developmental states concentrated economic decision-making in finance, industry and trade ministries. Yet, a developmental state with the characteristics just described might be compatible with the scientific procedures of EIA. After all, if states are concerned with broad, long-term national interests, they should consider the environmental foundations of their economic projects. The concept of sustainable development was coined to capture exactly that balance (Happaerts & Bruyninckx, 2014). The highly technical quality of scientific evidence and analysis is also compatible with the mode of analysis of the developmental state.

Including information and evidence on impacts from communities and social scientists is more difficult. From the standpoint of the classic developmental state and its proponents, the understanding of the role of society is a close match to the 'NIMBY' (Not in My Back Yard) understanding of local resistance to projects. In this perspective, such a response represents narrow self-interest in the face of public goods. Daniel Aldrich (2008, p. 1) summarises this point of view: 'Everyone wants cheap gasoline and lower heating bills, but no one wants to live next to an oil refinery or a fuel storage facility'. From a different perspective, empirical research on the NIMBY phenomenon has shown that such interpretations greatly understate the diversity and character of much opposition to projects (van der Horst, 2007). Citizen responses may even be constructive, offering real and manageable improvements to proposed projects. Decades of citizen resistance to hydroelectric dams in Brazil, for example, have forced project planners to be much more careful in their project designs than they were in an earlier generation of plants (Hochstetler, 2011).

Evaluating the effectiveness of EIA in the twenty-first century development strategies requires sensitivity to the multiple understandings that stakeholders may bring to the concept. An earlier generation of scholarship assumed that effective EIA meant matching a particular set of valued procedures, already described as the procedural approach. The growing proliferation of stakeholders requires attention to the standards they hold. Substantive effectiveness 'shows the achievement of expected objectives', usually sustainability (Chanchitpricha & Bond, 2013, p. 66–67). But other stakeholders may be particularly concerned with transactive effectiveness, which asks EIA to produce outcomes at low cost in an expeditious time frame. Finally, a normative lens on effectiveness considers broader societal and individual goals that are set by the context or culture. For example, these could include democratic norms that look for a fit to popular preferences. From the start, the larger number of stakeholders in modern developmentalism makes conflict over whether EIA is effective to seem more likely.

To move beyond these hypothetical discussions of the relationship between developmental strategies and EIA processes, the next section compares EIA processes in the Latin American countries. The comparison emphasises how EIA creates a framework that can be used to bring evidence about environmental and other impacts into economic policymaking. In this comparison, Brazil rises to the top, with procedures that appear among the strongest in the region.

EIA in Latin America

Table 1 presents the results of an index of the quality of the national EIA processes across the Latin American region. To construct this index, the author examined online documentation of the legislation and regulation of the most recently established EIA process in each country, along with any available secondary materials (for example, Abers, 2016; Barandiaran, 2015; Hochstetler, 2011; Jaskoski, 2014; Palerm & Aceves, 2004). Since, in most cases, only the legislation itself was available, the index stresses factors that can be identified in the text of EIA regulations. Scores were averaged for four subcomponents of the overall evaluation. The evaluations are set comparatively so that a high score means the EIA process is good for the region, but not necessarily an ideal EIA practice. As a rough indicator of procedural effectiveness, the index treats all components as equally important and is focused on national-level processes.

The first component of the index asked whether it is an environmental agency itself that assesses and approves the environmental impact study. Initial versions of EIA often placed responsibility for environmental permitting with the same sectoral ministries that were responsible for promoting particular economic activities. For example, a Ministry of Mines would issue permits for mining projects it was otherwise supporting, a clear conflict of interest (Jaskoski, 2014, p. 875). By the time of the latest iterations of EIA regulations, virtually all Latin American countries gave responsibility for environmental permitting to environmental agencies, a choice that is likely to improve consideration of environmental

Table 1. EIA ratings for Latin American countries.

	Score (of 3; 1 is highest)	Agency	Transparency	Consultation	Established
Argentina	3.0 Low	3	3	3	3
Bolivia	1.8 Medium	1	2	1	3
Brazil	1.3 High	1	1	2	1
Chile	1.8 Medium	1	2	2	3
Colombia	1.8 Medium	1	3	2	1
Costa Rica	2.0 Medium	1	2	2	3
Cuba	1.8 Medium	2	3	1	1
Dominican Rep.	2.0 Medium	1	2	2	3
Ecuador	1.5 High	1	3	1	1
El Salvador	1.5 High	1	2	2	1
Guatemala	2.0 Medium	1	2	2	3
Honduras	1.8 Medium	1	3	2	1
Mexico	1.5 High	1	2	2	1
Nicaragua	2.0 Medium	1	2	2	3
Panama	1.8 Medium	1	2	1	3
Paraguay	2.5 Low	3	2	2	3
Peru	2.5 Low	2	2	2	3
Uruguay	2.0 Medium	1	2	2	3

Source: Author's data based on review of EIA texts, as described in the article manuscript. Venezuela is excluded because of legal and regulatory uncertainty at the time of writing.

impacts. The exceptions are Argentina, which still had not regulated a 2005 law calling for a single national EIA process, Paraguay and Peru, which passed a law moving all EIAs to its environmental ministry but has still not provided regulation for the implementation of the law (Jaskoski, 2014).

A second component of the index evaluated the transparency of the process: how easily an interested party could have access to an environmental impact study. The availability of the EIA for study by members of the public, including interested scientists and academics, potentially promotes higher quality EIA studies, following the logic of studies on the role of transparency in policy-making (Gupta & Mason, 2014). When evidence is presented more broadly, additional actors can review it and suggest improvements and additional information. The prospect of such review should encourage the firm undertaking the study to produce a higher quality product in the first instance.

The variation in availability was substantial. At the high end, all the documents associated with Brazilian national licencing (since 2005) and some state licencing processes are publicly available on an easily searchable website. This can run to thousands of technical pages and represents high accessibility for knowledgeable Portuguese readers with Internet access (Hochstetler, 2011, p. 356). At the other extreme, in countries including Colombia and Ecuador, citizens must individually petition to see an EIA. The modal format in the region is that projects are announced in a public newspaper or sometimes the environmental agency's own communications and interested citizens then have between 10 and 20 days to go to a central location to see the study or a summary of it. Consequently, levels of transparency around impact assessment studies are generally quite low. Throughout the region, meaningful access to EIA documents is a special problem for speakers of indigenous languages (Palerm & Aceves, 2004).

Due to limited availability of information, it is not surprising that citizen participation is not central to EIA in most countries, the third component of the index. A few countries – Bolivia, Cuba, the Dominican Republic, Ecuador and Panama – require consultations with affected populations, and more do so for certain categories of high-impact projects. Bolivia and Peru have recently joined the small number of countries that require 'free prior informed consent' from indigenous populations for projects in their lands (Fontana & Grugel, 2016, p. 253). For high-impact projects, most countries have some provisions for possible public hearings, although there is a great deal of variance in how likely they are to happen. The most common provision for citizen participation is that citizens can submit written comments on the EIA that will be considered and possibly included in the final assessment. The EIA regulations are uniformly opaque on the details of how the consultation is to be done. Given the limited availability of information, consultation is often limited by citizens' lack of knowledge of the project beyond what is presented in the consultation meeting itself (Abers, 2016).

Finally, the study considered the length of time the country's EIA process had been in place, asking whether any major revisions have been made in EIA processes since 2002. The technical quality of environmental regulation implies that even a well-meaning and capable government needs years to put new requirements into practice (Landim & Sánchez, 2012). Impact assessment as widely practised requires double teams with substantial expertise: one in the private sector to generate the original assessment of the environmental impact and how to mitigate it, and one in the public sector to evaluate the private analysis (Glassen et al., 2012). The array of necessary skills crosses many disciplines. In addition to state capacity, the length of time regulations in place also affects firms' and citizens' familiarity with the process. Regional countries generally created environmental agencies and regulations only after undergoing transitions to democracy in the 1970s and 1980s, with Brazil and Mexico being the most prominent exceptions (Hochstetler, 2012).

Formal arrangements of the kind surveyed here provide a useful snapshot of the procedures regional countries have established to try to bring evidence about environmental impacts into their economic decision-making processes. These four dimensions yielded the EIA capacity measures in Table 1 and the final summaries of low, medium, and high procedural effectiveness. They show that most regional countries struggle to do EIA well: they have recently established procedures with limited transparency and consultation. Scientific expertise is also often limited. For countries with low EIA effectiveness, the return to developmentalism looks much like the classic version with little attention to the environmental and social impacts EIA is meant to spotlight. The higher ranked countries offer more opportunity to assess how policy-makers and others respond to newly systematic evidence about the impacts of their economic projects. The final section of the paper examines this question in Brazil.

EIA in Brazil: evaluations and responses to EIA practice

Brazil is one of a small number of Latin American countries that initiated its environmental institutions on a timetable of its own choosing, creating a national environmental agency in 1973. A National Environmental System followed in 1981, leading in turn to the second environmental licencing requirement in the region in 1986 (Hochstetler, 2012). The Brazilian EIA reflects longer experience and more qualified analysts than that of its Latin American neighbours. There have been steady improvements, although its licencing still shows many gaps in quality (Landim & Sánchez, 2012). The Brazilian Institute of the Environment (Ibama, in its Portuguese acronym) in the national Ministry of the Environment evaluates multistate and complex projects, while state licencing agencies handle smaller projects. Only the national level is discussed here.

Most countries have just one environmental licence, while Brazil requires three, for the planning, construction and operation stages. The licencing agencies often put numerous conditions on the licence of one phase that must be addressed before the next can be granted. Ibama and the subnational agencies handle not just physical environmental impact but also include determinations from other agencies about impacts on indigenous groups, national historical patrimony and other impacts that become part of the final licencing decision. Their own EIAs also address social and economic impacts, so Brazil's 'environmental' licences cover an unusually broad terrain. Evidence from the social sciences is particularly important for evaluating the non-physical impacts, as is consultation with local communities.

In 2003, the Workers' Party won the national presidency and, over the next years, launched expansive, state-led programmes of investment in infrastructure (Montero, 2014). This expansion coincided with unusually high prices for commodities that also spurred economic expansion. All these projects needed environmental licences, with the number of licencing processes under review leaping from 791 in 2005 to 1672 in 2013. In 2012 alone, Ibama issued 700 licences (Moura, 2016, p. 125). Environmental licencing became the front-page news as the over-burdened system struggled under an avalanche of criticisms that it was either far too rigid and time-consuming or far too permissive, with corresponding proposals for reform. Licencing agents also sought improvements in a system they acknowledged to be flawed. The remainder of this section explores those different proposals and how they conceived of the effectiveness of Brazilian EIA.

Critiques from inside

Licencing practitioners agree more readily on the flaws than the positive elements of Brazil's environmental impact system. In a recent survey, 77% thought implementation requirements were poorly executed, 77% criticised the participation of non-environmental government agencies and 76% saw problems with public participation. Almost as many saw a misfit between EIA and guidelines in other planning processes (74%) and a failure to learn lessons (73%) (Duarte, Dibo, Siqueira-Gay, & Sánchez, 2017, p. 4). Many of these complaints are about actors outside the licencing agencies themselves, which saw a number of improvements during this time. In 2003, Ibama had just 3–4 permanent licencing analysts and about 100 consultants (Ibama, 2009), but this was followed by a substantial professionalisation of the agency. The number of federal licencing analysts had grown to 400–450 a decade later, all of whom had passed qualifying tests in various disciplines. Retention has been an ongoing issue, however, partly due to low salaries but also to a sense that the career is not socially valued (Toledo, personal communication, September 28, 2014).

The other feature of these critiques is that they are primarily procedural in nature, as are the solutions that licencing practitioners have proposed. For example, in 2011, they contributed to a new legislation that set out the responsibilities between federal, state, and municipal levels of licencing more clearly (Law 140/2011). This law did little to change licencing behaviour, but the topic had been a major target of judicial challenges, which now changed the focus (Toledo, personal communication, September 28, 2014). The state licencing agencies and the National Environmental Council (Conama) also worked together with hosting communities and the wind power industry to develop new guidelines that would allow wind plants that met certain criteria – not on dunes or in protected areas, with community consent and so on – to receive expedited licencing (Conama Resolution 462/2014). Part of the motivation for this was to use environmental licencing requirements less as a block to undesirable projects and more as an incentive for choosing greener alternatives (Santos, 2014). This response reflects a significant and constructive effort to rethink the purpose and framing of environmental licencing.

The Ministry of Environment has also been working in recent years to try to develop a more comprehensive new national legislation on licencing, but environmental activists have withdrawn from what they call an excessively hasty process, while business and economic actors have been working toward very different alternatives. In general, the proposals for change from the licencing agencies themselves are ones that value the evidence that EIA produces, but seek judiciously to streamline the process where possible.

Critiques from policy-makers

This approach is too cautious for many of Brazil's politicians and proponents of new infrastructure and extraction projects. They have come to see environmental licencing as a significant bottleneck in national development plans (Bragagnolo, Lemos, Ladle, & Pellin, 2017; Knijnic, personal communication, September 23, 2014). Legislators proposed at least 15 versions of laws governing environmental licencing between 2004 and 2015, according to a background study conducted for the lower house of the National Congress (Hofmann & de Araujo, 2015, pp. 3–4). Many aimed to rework environmental licencing altogether, while others took on only pieces of the process. Several were intended to make the process more rigorous, but the proposals that advanced furthest were designed to speed up and minimise the intrusion of EIA.

Legislative Proposal 3719 originated in the lower house in 2004 and shares important similarities with the Senate's Legislative Proposal 654 of 2015. Both are associated with parties that were once part of the governing coalition with the Workers' Party (PT) but led successful impeachment proceedings against the PT president Dilma Rousseff in 2015. The heart of their economic base is the 'ruralist' coalition of industrial agriculture and natural resource extractors, but key industrial coalitions have also signed on to similar proposals (Confederação Nacional da Indústria, 2013). Both proposals aim to streamline the process by exempting more activities from licencing, particularly those that were central to the state's developmentalist agenda. They also set short time frames for evaluation, limited the ability of assessors to ask for more information and minimised public input. PLS 654 also shortens the process to have only one licence, while PL 3729 removes criminal punishment for infractions. The net result would be to reduce significantly the amount of evidence gathered and analysed about environmental impacts.

As both laws moved through congressional hearings and committee votes in 2016, environmental licencing professionals, activists and legal representatives presented counterarguments in testimonies and collective statements. Their critiques claimed that the proposed legislation would destroy effective licencing to an extent that would be unconstitutional and ineffective. One group of 136 organisations signed a Manifesto against PLS 654, arguing that 'To eliminate the spaces for direct participation by affected and interested people is the least efficient way of facing the conflicts inherent in big infrastructure projects' (Instituto Socioambiental, 2016). If communities could not participate normally, they threatened they would judicialise the claims.

The backdrop here is that Brazil has a Ministério Público (MP) (loosely, a public prosecutor) who is very active on environmental and consumer issues and frequently focuses on EIA cases (McAllister, 2008). Interestingly, one of Ibama's reactions to constant legal oversight was to make as much information publicly available as early as possible, in hopes of learning of potential opposition before it could move to a formal complaint process (Ibama, 2009). All actors carry out the licencing process fully aware that their decisions are likely to be closely scrutinised and may end up in court (Knijnik, 2014; Toledo, personal communication, September 28, 2014). Court injunctions can stop a project altogether and more frequently delay them in ways that make the project less viable (McAllister, 2008). The Federal MP presented a legal statement (Parecer Jurídico No. 4–4ª CCR) in 2015 stating its legal objections to the Senate legislative proposal, making it clear it would challenge the proposal and similar ones in court as unconstitutional.

A similar set of partisan actors was working simultaneously to push a constitutional amendment (PEC 65/2012) that would have forestalled such constitutional challenges. Clumsily written, it asked only that an environmental impact study be presented for priority economic projects, but did not provide for analysis of the study, and said the project could not be blocked. It also omitted citizen consultation. The Federal MP again presented an analysis saying that the amendment conflicted with too many immutable constitutional provisions, such as human rights and environmental protection provisions, to be allowed (Ministério Público Federal, 2016). This particular version now appears dead. The sentiments and motivations behind these initiatives remain unchanged, however, and similar proposals are likely to return.

This group of stakeholders is consistently most concerned with the transactive effectiveness of EIA, looking for low costs and short time frames. Its proposals show much more concern with the developmental effects of projects, and openly advocate setting aside sustainability criteria for especially important projects, much as classic developmentalism did.

Critiques from civil society

Even as one group of actors is looking to accelerate environmental licencing and limit the evidence it collects and uses, another is working to expand the scope of Brazilian EIA. The title of a 2014 book (Zhouri & Valencio, 2014) (translated) – *Forms of killing, dying, and resisting: The limits of negotiated resolutions of environmental conflicts* – succinctly captures the stakes considered to be at risk. EIA and its consultations are exactly such a 'negotiated resolution', now increasingly rejected by activists who would like to see a much more substantial debate about the substantive socio-environmental costs of projects than environmental licencing has allowed so far.

Consultations with affected communities are one of the weakest parts of Brazilian EIA, according to a survey of 131 academic papers published in English and Portuguese on the topic. Evaluation of socio-economic impacts is seen to be frequently ignored or highly subjective (Duarte, Dibo, & Sánchez, 2017, pp. 275–276). Consultations for the Belo Monte hydroelectric dam included 4417 participants, who asked 682 questions in hearings that ran for 30 hours (Hochstetler, 2011, p. 360). Even so, indigenous and other local forest peoples were still heavily affected by the final project, and opponents supported a series of legal challenges in Brazil and at the Organization of American States that targeted the procedures and outcomes of the EIA process (Hochstetler, 2011). In six routine infrastructure cases since 2000, Abers (2016) found that the consultations were perfunctory, with most of the focus on presenting the project. Questions were answered in highly technical language or brushed off, and there was little evidence that the community participation affected the projects. Communities have been able to use the legal system to support their claims (Hochstetler, 2011), but it requires framing the arguments primarily in procedural terms.

For activists and social scientists alike, one attempted solution to this flaw has been gathering more evidence about socio-economic and socio-environmental impacts, even developing their own alternative impact studies (McCormick, 2009). But their aim is increasingly one that may challenge the institution of licencing itself. Inside Ibama – and in the world of EIA more generally – environmental licencing is understood as a way to improve a proposed economic project from the environmental point of view rather than a decision about doing the project or not (Glassen et al., 2012; Ibama, 2009; Toledo, personal communication, September 28, 2014). In the view of many environmental and community activists, EIA should be about the latter decision, with communities given much more ability to reject projects they oppose and some categories of activities like large hydroelectric dams excluded altogether (for example Greenpeace, 2016). This is exactly the outcome that the project proponents want to make impossible.

Conflicts over the right way to think about effectiveness underlie such differences. Economic actors favour transactive effectiveness, often directly traded-off with the substantive effectiveness that activists demand. The two sides have been engaged in open conflict over which of them can claim normative effectiveness, or the right to say their view reflects the popular will. Licencing practitioners prefer to take a narrower view and focus on procedures and information as they relate to particular projects, without engaging the broader claims.

Conclusion

EIA is an important regulatory instrument for bringing evidence about environmental and social impacts to bear on policy-making. In principle, EIA should make economic projects more sustainable, but it will only do so if evidence is collected well and analysed carefully. This is a challenging task for developing countries, especially when states are setting out ambitious development strategies with extensive state participation. EIA effectively puts one part of the state against another, a fight that economic agencies typically win. The article shows that current EIA frameworks across the Latin American region are set up with characteristics that limit the likelihood that EIA processes will collect, disseminate and analyse socio-environmental information in influential ways. Brazil emerges as one of the regional bests – on paper – but a final case study shows that, even there, EIA is critically embattled.

Comparatively strong environmental licencing, with extensive judicialisation of important projects, has resulted in ongoing efforts to roll back EIA processes. Many policymakers would rather not have effective investigation of the effects of strategic projects or would rather not have more evidence than that which can be quickly gathered and allow the project to proceed. They are countered by civil society actors who are determined to hold the line, if not to push impact assessment further, with more attention to social data. The quiet improvements within the licencing community itself go largely unnoticed by both sides.

The documentation in Table 1 shows that the current round of developmentalism is different in Latin America: all the countries in the region have introduced regulations requiring procedures that will provide information about environmental and social impacts. However, some of the regulations themselves are too weak to achieve much. Even the country which has some of the strongest EIA procedures on paper, Brazil, shows that while EIA is crucial for inviting stakeholders who will speak for substantive sustainability ends, it is an invitation to a struggle rather than the gateway to a certain, balanced outcome between environment and development.

Acknowledgements

The author would like to thank the editors and reviewers as well as Javiera Barandiaran, Nitsan Chorev, Maiah Jaskoski, Eduardo Silva and Paul Steinberg for helpful suggestions on earlier versions of this manuscript.

Disclosure statement

No potential conflict of interest was reported by the author.

Funding

This article is supported by the Social Sciences and Humanities Research Council of Canada (Canada)

Notes on contributor

Kathryn Hochstetler is Professor of International Development at the London School of Economics and Political Science. She has published extensively on environmental politics in Brazil and Latin America, including the prize-winning *Greening Brazil: Environmental Activism in State and Society* (Duke University Press, with Margaret E. Keck).

ORCID

Kathryn Hochstetler http://orcid.org/0000-0003-2960-058X

References

Abers, R. (2016). *Conflitos, mobilizações e a participação institucionalizada: A relação com a sociedade civil na construção de grandes obras de infraestrutura)* [Conflicts, mobilisations and institutionalized participation: The relation with civil society in the construction of large infrastructure projects]. Brasília: IPEA, Discussion text.

Aldrich, D. P. (2008). *Site fights: Divisive facilities and civil society in Japan and the west.* Ithaca, NY: Cornell University Press.

Barandiaran, J. (2015). Chile's environmental assessments: Contested knowledge in an emerging democracy. *Science as Culture, 24*(3), 251–275.

Bragagnolo, C., Lemos, C. C., Ladle, R. J., & Pellin, A. (2017). Streamlining or sidestepping? Political pressure to revise environmental licensing and EIA in Brazil. *Environmental Impact Assessment Review, 65*, 86–90.

Chanchitpricha, C., & Bond, A. (2013). Conceptualising the effectiveness of impact assessment processes. *Environmental Impact Assessment Review, 43*, 65–72.

Confederação Nacional da Industria [National Industrial Confederation]. (2013). *Proposta da indústria para o aprimoramento do licenciamento ambiental* [Industrial proposal for improving environmental licensing]. Retrieved from http://www.ibram.org.br/sites/1300/1382/00003693.pdf

Duarte, C. G., Dibo, A. P. A., & Sánchez, L. E. (2017). What does the academic research say about impact assessment and environmental licensing in Brazil? *Ambiente & Sociedade [Environment & Society], 20*(1), 261–292.

Duarte, C. G., Dibo, A. P. A., Siqueira-Gay, J., & Sánchez, L. E. (2017). Practitioners' perceptions of the Brazilian environmental impact system: Results from a survey. *Impact Assessment and Project Appraisal*, doi:10.1080/14615517.2017.1322813o

Fontana, L. B., & Grugel, J. (2016). The politics of indigenous participation through 'free prior informed consent': Reflections from the Bolivian case. *World Development, 77*, 249–261.

Glassen, J., Therivel, R., & Chadwick, A. (2012). *Introduction to environmental impact assessment* (4th ed). New York, NY: Routledge.

Greenpeace. (2016). *Hidréletricas na Amazônia: um mau negocio para o Brasil e para o mundo* [Hidroelectric dams in the Amazon: Bad business for Brazil and the world]. Retrieved from http://www.greenpeace.org/brasil/pt/Noticias/Hidreletricas-na-Amazonia-um-mau-negocio-para-o-Brasil-e-para-o-mundo/

Gupta, A., & Mason, M. Eds. (2014). *A transparency turn in global environmental governance.* Cambridge: MIT Press.

Happaerts, S., & Bruyninckx, H. (2014). Sustainable development: The institutionalization of a contested policy concept. In M. Betsill, K. Hochstetler, & D. Stevis (Eds.), *Advances in international environmental politics* (2nd ed.). (pp. 300–327). Houndmills: Palgrave.

Hochstetler, K. (2011). The politics of environmental licensing: Energy projects of the past and future in Brazil. *Studies in Comparative International Development, 46*(4), 349–371.
Hochstetler, K. (2012). Democracy and the environment in Latin America and Eastern Europe. In P. Steinberg, & S. VanDeveer (Eds.), *Comparative environmental politics: Theory, practice, and prospects* (pp. 199–229). Cambridge: MIT Press.
Hochstetler, K., & Tranjan, J. R. (2016). Environment and consultation in the Brazilian democratic developmental state. *Comparative Politics, 48*(4), 497–516.
Hofmann, R. M., & de Araujo, S. M. V. G. (2015). *Desafios do licenciamento ambiental* [Challenges of environmental licensing]. Brasília: Câmara dos Deputados Consultoria Legislativa, Estudo Setembro 2015 [Legislative consultant, Chamber of Deputies, September 2015 study].
van der Horst Dan. (2007). NIMBY or not? Exploring the relevance of location and the politics of voiced opinions in renewable energy siting controversies. *Energy Policy, 35*, 2705–2714.
IBAMA. (2009). Confidential interview with former environmental licensing official of Ibama, Brasília, 2009.
Instituto Socioambiental [Socioenvironmental Institute]. (2016). *Manifesto contra o PLS654* [Manifesto against PLS654]. Retrieved from https://www.socioambiental.org/sites/blog.socioambiental.org/files/nsa/arquivos/manifesto_contra_o_pls_654_-_com_assinaturas_-_versao_final.pdf
Jaskoski, M. (2014). Environmental licensing and conflict in Peru's mining sector: A path-dependent analysis. *World Development, 64*, 873–883.
Johnson, C. (1982). *MITI and the Japanese miracle*. Stanford, CA: Stanford University Press.
Knijnik, C. (2014). Interview, director of the energy program of the Program of Growth Acceleration (PAC), Ministry of Planning, Brasília, September 2014.
Landim, S. N. T., & Sánchez, L. E. (2012). The contents and scope of environmental impact statements: How do they evolve over time? *Impact Assessment and Project Appraisal, 30*(4), 217–228.
McAllister, L. (2008). *Making law matter: Environmental protection and legal institutions in Brazil*. Stanford, CA: Stanford University Press.
McCormick, S. (2009). *Mobilizing science: Movements, participation and the remaking of knowledge*. Philadelphia, PA: Temple University Press.
Ministério Público Federal [Federal Public Ministry]. (2016). *Nota Técnica a PEC 65/2012 e as cláusulas pétreas [Technical Note on PEC 65/2012 and the immutable clauses]*. Brasília: Ministério Público Federal.
Montero, A. P. (2014). *Brazil: Reversal of fortune*. Cambridge: Polity Press.
Morgan, R. K. (2012). Environmental impact assessment: The state of the art. *Impact Assessment and Project Appraisal, 30*(1), 5–14.
Moura, A. M. M. de. (2016). Aplicação dos instrumentos de política ambiental no Brasil: Avanços e desafios [Applying the instruments of environmental policy in Brazil: Advances and challenges]. In A. M. M. de Moura (Ed.), *Governança Ambiental no Brasil: Instituições, Atores e Políticas Públicas* [Environmental governance in Brazil: Institutions, actors, and public policies] (pp. 111–145). Brasília: IPEA.
Palerm, J., & Aceves, C. (2004). Environmental impact assessment in Mexico: An analysis from a 'consolidating democracy' perspective. *Impact Assessment and Project Appraisal, 22*(2), 99–108.
Santos, J. P. F. (2014). Interview, Director of Department of Support for Conama, Brasília, September.
Wade, R. (1990). *Governing the market: Economic theory and the role of government in East Asian industrialization*. Princeton, NJ: Princeton University Press.
Weyland, K., Madrid, R. L., & Hunter, W. (Eds.). (2010). *Leftist governments in Latin America: Successes and shortcomings*. Cambridge: Cambridge University Press.
Zhouri, A., & Valencio, N. (Eds.). (2014). *Formas de matar, de morrer e de resistir: Limites da resolução negociada de conflitos ambientais* [Forms of killing, dying and resisting: Limits to negotiated solutions of environmental conflicts]. Belo Horizonte: Editora UFMG.

OPEN ACCESS

Low carbon energy and international development: from research impact to policymaking

Ed Brown, Ben Campbell, Jon Cloke, Long Seng To, Britta Turner and Alistair Wray

ABSTRACT

Few areas of international development research have seen as much transformation over recent years as those relating to energy access and low carbon transitions. New policy initiatives, technological innovations and business models have radically transformed the configuration and dynamics of the sector, driven by the urgency of ongoing climate change. This article asks how, given these rapidly moving contexts, policymakers can engage with research at different scales to gather evidence needed for effective decision-making, particularly within the context of the frequently opposing aims of increasing energy access and climate change mitigation. The authors trace the general debates around how research impact is conceived within different constituencies, before exploring the relationship between policymakers, the academic community and other stakeholders within the specific context of energy and international development research. Drawing on cross-cutting lessons from thirteen research projects funded by UK research councils and government under the Understanding Sustainable Energy Solutions programme, they examine critically ways in which impact and engagement have been conceived by both researchers and research funders. They ask how those lessons can feed into the design of future initiatives to make low carbon transitions meaningful as pathways for inclusive development in communities in Africa and Asia.

Introduction

Climate-change-driven low carbon energy transitions have become an increasingly prominent component in visions for sustainable development over the past decade (Brown, Cloke, Gent, Johnson, & Hill, 2014; Ockwell & Byrne, 2016). The insistent emergence of climate change in fields of governance (frequently in the face of fierce resistance) is gradually forcing the development of new ways of 'doing policy' as policymakers grapple with meeting carbon emissions targets at the same time as global temperature rises are heralding new and forbidding climatological phenomena. Some of the challenges of increasing energy access within the context of low carbon energy transitions are captured in outline form in the 'Energy trilemma' conceptualised by the World Energy Council (2013). Developing a policy that addresses the three core trilemma dimensions of energy security, energy equity and reducing energy's GHG emissions demands recognition of the complex interwoven links between public and private actors, governments and regulators, economic and social factors, natural resources and the pressing urgency of global warming. Although not without its limitations, the trilemma concept has the potential to provide a multi-faceted integrating framework for discussing UK government and international energy sector futures, in a way that complements the wider climate change debate while emphasising the role of policymakers.

Access to modern energy services has long been considered as a key contributor to sustainable development, although we should be cautious about drawing too simplistic parallels between energy access and livelihood improvement (Cook, 2011). The complexity and multiple dimensions of energy equity create difficulties in disentangling energy poverty from the two other elements of the trilemma: climate change and energy security (Tomei & Gent, 2015). Despite this complexity, an emerging consensus on the need to tackle global energy poverty in an unprecedented way has catalysed a step change in energy access activity since the launch of the UN Sustainable Energy for All initiative in 2012 and subsequent agreement on the Sustainable Development Goals (SDGs), including the SDG 7 to ensure access to affordable, reliable, sustainable and modern energy for all by 2030. This consensus has resulted in a unifying sector framework for countries and development partners alike, for tracking progress against the ambitious international goals for energy access and, most important for our purposes, evaluating the impacts of research designed to support these initiatives.

Nevertheless, innovative but untested visions of energy futures at the national and international scale must also be underpinned by sound reasoning and critical assessment of the evidence on the role low carbon energy innovation can play in livelihood options at the local scale, particularly in places and social contexts of apparent development need. Communities of the poor, who thus far have not been well served by fossil fuel-driven models of development, need to be able to appropriate, assume and apply low carbon solutions for problems in their home contexts. The literature on technology and development (Bray, 2007; Dove & Kammen, 2015; Feldman & Briggs, 2012; Leach, Scoones, & Wynne, 2005) is, however, full of examples where techno-social visions deploying inappropriate equipment, inputs and infrastructures have followed naïve paradigms of innovation. Social scientists have analysed ways in which interventions (by energy companies, NGOs or governments) and policies have missed research messages about the need to understand diverse contexts of users, resulting in top-down engineered solutions failing to connect with local needs, preferences and aspirations, not to mention gender norms working against women's energy interests, practical dispositions and aspirations.

Socio-technically aware research and project implementation methodologies have the great potential to reduce such failures. In terms of participatory low carbon transitions, knowing where to look for impact and understanding the formal/informal institutions of everyday coping mechanisms of fuel and energy service flows would greatly enrich the field of low carbon energy innovation in the Global South; the granular qualities of peoples' needs, preferences and means for taking action would be made much more visible (Bazilian & Pielke, 2013; Ockwell & Byrne, 2016; Rolffs, Ockwell, & Byrne, 2015; Winther, 2015). As a set of guidelines, Agenda 21 of the Rio Earth Summit formalised a paradigm change for sustainable natural resource management systems in the development industry's adoption of sustainability. One message was that natural resources can often best be managed by self-organising local systems of surveillance and cautious harvesting, which applies equally to low carbon energy.

How, though, can such impact-focused research initiatives be best promoted? Effective sustainable energy impact is located in an ecology of intra-academic practices that both research funders and researchers themselves recognise must prioritise the interface with non-academics, the use of policy discourse, local knowledge and critical poverty analysis, to make a difference and share the learning process. Here, we explore the approach of the UK Department for International Development (DfID) in initiating impact-focused research in this field via support for a programme of collaborative work undertaken through the Low Carbon Energy for Development Network (LCEDN) since 2013, designed to network and strengthen the UK-funded research being undertaken in support of pro-poor renewable energy interventions. The article proceeds by exploring some elements of the current wider debates over the meaning and measurement of research impact, before going on to explore both the evolution of an integrated programme of UK-funded sustainable energy-focused projects specifically aimed at fostering that ecology, and how the ongoing learning process from those experiences is being used to feed the direction of further research interventions.

Transdisciplinary research and impact

Since 2000, a significant shift has occurred within the evaluation of the quality of academic research to ensure such assessments encompass the impact of that research on policy outcomes. In the UK, for example, a considerable debate (Reale et al., 2017; Strathern, 2000) has arisen about the measurement of impact, and tensions certainly exist between the understandings of impact held by research councils, academics and other stakeholders such as government departments, businesses and civil society organisations. One central element of this debate is the distinction between linear and networked models of how impact might be achieved. The more linear models of how academic research produces an impact, which have influenced how research is measured through the UK's research excellence framework (REF) and other similar exercises, are coming under increasing critical analysis (Nicholls, 2015; Shortt, Pearce, Mitchell, & Smith, 2016). Here we take the stance that it is a networked, rather than linear, model of the impact that needs to be embraced and deployed. A networked approach to the impacts of research in low carbon energy transitions conceives impact to be dialogically produced as an outcome of exchanges of viewpoints, thereby recognising a diversity of positions and perspectives that result in more than the sum of aggregate parts and achieving a means of moving beyond interdisciplinary lines. The impact is an effect of interaction, networking by mutual learning and ways of seeing transitions from different angles. In this respect, innovative methodologies such as the participatory conception of impact pathways take cognisance of mutual understanding and interactions that facilitate innovation (Alvarez et al., 2010). Co-production of socio-technical innovation means livelihood adaptations are taken up for distinctive purposes and conditions, and in relationships of differential power. Here, the field of operation for creating a dialogue about impact is what matters.

By contrast, the competitive model to which the UK higher education (HE) sector is subjected works in the opposite direction: from Personal Development Reviews for individual academics and researchers, through research groups, centres, departments and schools, collaborative effort is sidelined in favour of individualistic 'silo-ised' metrics that are easier to measure, and which can be used to promote the overall performance of the university in the pursuit of funding. Although UK HE institutions frequently discuss transdisciplinary approaches as vital to research and teaching, current HE financial models and the drive for inter-institutional competition effectively discourages such initiatives, even though recent reports on the HE sector continue to stress the importance (and lack of) such initiatives (British Academy, 2016).

We, therefore, 'need to talk about impact' (Smith & Stewart, 2017, p. 109) within the low carbon energy for development community in a way that makes sense not only across the variety of actors involved, who come from extremely different backgrounds in terms of their disciplinary readiness to engage with policy, but also in how such metrics for impact requiring transdisciplinarity can be harmonised with institutional pressures that stand to frustrate them. This approach means thinking about pluralising understandings of impact rather than engineering them in advance to predefined ends. Robbins, Wield, and Wilson (2017) discuss their mapping of research impact in development projects, and initial expectations in their thinking that pro-poor humanitarian and small-scale 'frugal innovations' would have been prominent in the UK REF impact case studies that they scrutinised. Instead, they found a prevailing discourse of ecological modernisation in which economic outcomes and large scales were most valued. The authors highlight that this was a choice made by the case study authors about the kind of storylines of impact they considered would be favourably received by those charged with 'judging' impact. Academics with more humanitarian inclinations and less techno-centric impact narratives are not intrinsically opposed to the more linear definitions currently shaping research impact submissions, but Robbins et al. (2017) found little discussion on the differences between 'academic' impact and meeting development policy objectives.

Evolving approaches to development and technology innovation

Current perspectives on the uptake of renewable energy technologies across the Global South can sometimes seem extremely linear and deterministic in their assumptions about research impacts. The goal of much research, it would appear, is to design technologies that can deliver electricity at scale and at relatively low cost, without increasing global carbon emissions. The focus is generally on the more easily measurable goal of electricity access, despite the massive need for improvement in other key elements of energy provision such as heat for cooking or fuels for transport and motive power. Impact in this context equals the number of sales of equipment or the numbers of connections made to networks, rather than the implications of that access upon livelihood opportunities, the quality of life and pre-existing inequalities and social divisions.

This emphasis has tended to feed the dominance of perspectives that prioritise over-simplistic promotion of North–South technology transfer and technological determinism in the evolution of policy. Other areas of research, however, have had a stronger emphasis on the valuing of local knowledge practices, thereby inspiring an immensely rich literature on local and indigenous capacities for learning and adapting to a wide range of different techno-economic transformations (Leach et al., 2005; Richards, 1985). This literature advocates symmetry between the knowledge of 'scientific' experts from Northern countries and the homegrown empirical knowledge of local people. Critical development studies interrogate how the 'social slot' gets left to pick up the user interface at the 'end of the pipe', in contrast to commonplace notions of interdisciplinarity, in which knowledge is imagined to be capable of selectively recombining knowledge from different component domains of natural and social science (Wellberry, 2009).

In contrast to piecing different parts of the jigsaw together with compatible data sets, the different framings of problems and perspectives offered by considering all actors in a social field as plausible innovation stakeholders has opened up development policy and intervention paradigms, revealing the differential benefits to broad categories of people and options available for innovation, for instance by foregrounding questions about gender and energy (Winter, Matinga, Ulsrud, & Standal, 2017). These types of impact do not lend themselves to additive logics or to linear measures of progress or impact, but to more normative evaluations of fluid ways of life and their potential for change (Bulkeley & Castan Broto, 2013; Castree et al., 2014).

Renewable energy project beneficiaries are not just receivers of technology; they are reflective observers of a complex, and challenging world, who prioritise time and resources in constrained circumstances. Individual or communal decisions about whether to adopt a new solar home system, wind turbine or biogas unit will be influenced, among others, by observation of the technology's effectiveness for other people they know. Low carbon development in off-grid distributed contexts of use needs to pay particular attention to organic kinds of technology governance practice, and to understand the relationships that affect uptake, capacity development and impact (Byrne et al., 2012).

In addition, the impact of research depends not only on how research findings are communicated but also on the receptivity of different actors to hearing a given message, interpreting a case study or to finding a piece of analysis convincing. Actor networks are key, and not simply the human protagonists: equipment and installations of infrastructure have different kinds of impact for different sets of users. Low carbon energy technologies should be seen as assemblages with possibilities for applications not foreseen by their designers (de Laet & Mol, 2000). They carry capacities for modification in an arrangement, by which actors may find them more or differently useful; and their impacts depend on how different actors can align them with domestic, productive and aesthetic purposes. As a direct result, the impact can be tracked in changes made to people's ability to see value in realistic enhancements to their lives, not in unreasonable transformations of technical possibility. Moving beyond collaborative forms of interdisciplinarity, problem-setting and innovatory capacities of beneficiary communities and other stakeholders makes a vital contribution and steers the technological input to development projects through degrees of ownership. Local applications and productive utility of materials and purposes align themselves into formulations that can be borrowed and tested in other locations. Bespoke solutions for context-specific needs make excellent case studies for thinking about scaling up, as illustrated by the programmes described below, which attempt to piece together narratives about such cases. The 'inclusivity profile' of particular technology deployments, for instance, needs to be understood through their life cycle of take-up and potential for diffusion. In their work on Participatory Impact Pathways Assessment, Alvarez et al. (2010, p. 956) underline that: 'The extent to which the good ideas and innovations that [project participants] are generating influence, and are influenced by, other actors depends upon how people are linked to each other, the nature of those linkages, local norms, and power relationships.'

As a vital corollary to technology deployment, one key component in bringing sustainable energy to marginalised and poor communities in the Global South is correlating energy access with other vectors of poverty to compare relative capacities for impact in 'last mile' innovation. This revised way of considering impact deploying recognisable, measurable and comparable kinds of information depends on robust sets of categories with enduring value for use by a range of actors, public, private, third sector and regular citizens. Organising information about the impact in this way can allow for both greater research accountability and project implementation to reach out to targeted groups of beneficiaries, while also achieving effects of redistributive and procedural justice with regard to energy access outcomes.

Evolving priorities and policy drivers in the UK context

The UKAid Economic Development Strategy (DfID, 2017) is the clearest pronouncement by the UK government to date attributing a central role to energy provision in delivering inclusive growth in UK partner countries, through climate-smart approaches, innovative financing and working with other UK government departments as part of a coherent development strategy (DfID, 2017, Chapter 3). The strategy is intended to provide the basis for an evolving framework for UK energy policy support encompassing interlinked objectives of ensuring energy access for the poor, meeting the needs of energy for economic development and enhancing environmental sustainability, while also recognising that delivery should be context-aligned. This positive endorsement of the key importance of promoting transitions towards greener energy strategies among UK partners seems, however, to stand in sharp contradiction to other areas of UK international policy. For example, UK Export Finance continued provision of substantial funding for fossil fuel technology exports, which is being undertaken with little or no reference to this strategy and at a level far higher than that provided for renewable energy exports (see Makhijani, 2014).

Outside the political economy context of individual nation-state decision-making processes, global energy sector development continues to be driven strongly by the falling prices of renewables, increasing competitiveness and development of renewable energy markets, and the growing use of smart technologies, mobile-enabled payments, innovative financing and remote data monitoring. The interests of a range of different key stakeholders (public and private sector, donors and investors), and energy provision options involved in this rapidly evolving complex of interests, demand that more attention is paid to a whole-energy system approach to drive coherence, policy change and commercial viability to deliver the energy access goals.

DfID has expanded investment not only in energy programming, but also in energy research (see Figure 1). UK clean energy research for development has grown dramatically since 2010 (UKCDS, 2014), in part at least reflecting the development priority now placed on energy, the challenges and opportunities that achieving universal access to modern energy services presents, and the goals outlined above, not to mention the business possibilities for UK firms. What, though, have DfID's intentions been in terms of the perceived

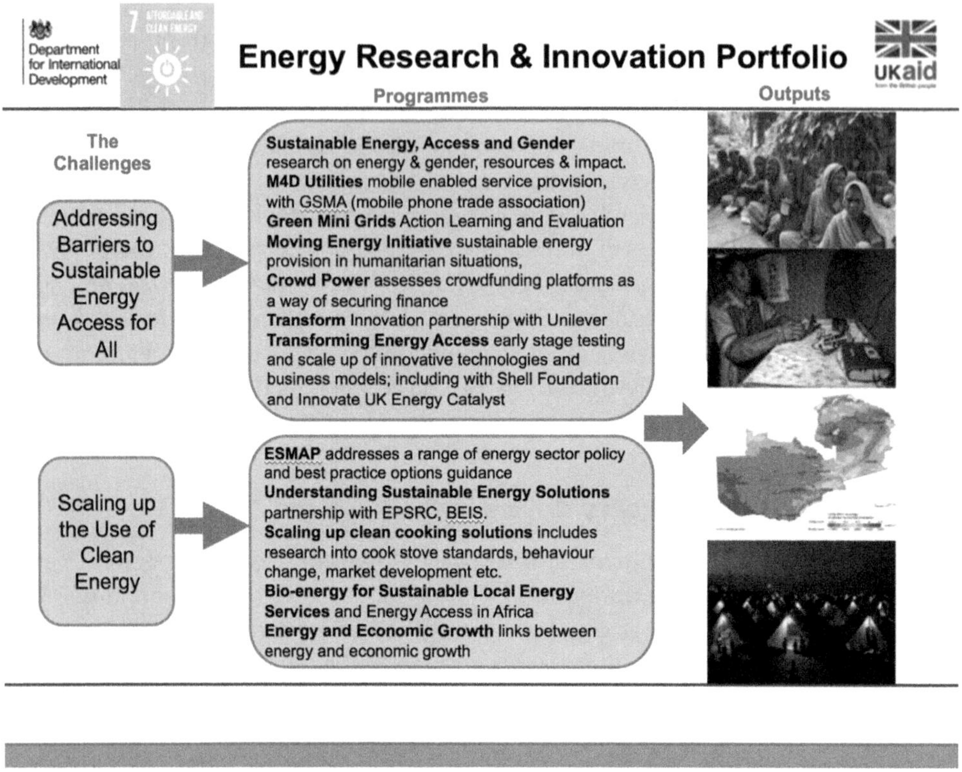

Figure 1. DfID energy research and innovation portfolio 2017.
Source: (supplied by Alistair Wray).

impacts of that research investment, the type of evidence that it is supposed to generate and the research partnerships supported, and what have been the actual effects?

A recent review of DfID (2016a) research considers the contribution that science and technology can make for development, the financial resources available in support of research and the importance of high impact research, innovation and strategic partnerships, but says little about the role of the social sciences. In part, this mirrors the UK HE and research community, which has only recently begun to map out and understand the indispensable role that collaboration between the social and natural sciences has in pushing forward the boundaries of low carbon energy transitions, and in rebalancing the prevalent voices in framing energy transitions discourse. DfID plans to continue to invest 3 per cent of the aid budget on research, judging value for money in terms of impact, additionality, quality, deliverability and cost, and it is the central role of the research and practitioner community in the UK to make the case for techno-social collaborations in this area. An evolving, multi-layered and inclusive understanding of impact is therefore key, requiring strong techno-social partnerships in priority research areas to optimise impact, and emphasising that outcomes should be judged on impact for priority stakeholders rather than purely on academic merits. It needs to be acknowledged furthermore that the projects referenced subsequently in this article are but one arc of an energy transitions wave.

Transforming and scaling up smarter energy access is an integral part of the DfID research agenda and climate and energy research and innovation are due to continue to receive significant investment. At the time of writing, the UK had committed to doubling renewable energy research, development and demonstration over 5 years by 2020/2021 under the Mission Innovation (2017) initiative and the UK Energy Innovation Programme. As part of an internal review of clean energy research and innovation needs, DfID (2014) identified five main drivers, which have helped shape the resulting portfolio after 2014:

- International development priorities in promoting sustainable energy for all;
- Growth in demand for energy, supply response scenarios and climate-resilient considerations;
- Emerging policy positions – fossil fuel subsidies, energy security, diversity of supply and energy efficiency;
- Raising awareness of renewable energy resources, their real costs and opportunities;
- Promoting technology innovation and the role of the private sector.

An informed policy analysis of the development challenges, review of research objectives and analysis of development potential has led to the emergence of three inter-linked research themes: addressing the barriers to sustainable energy access for all; scaling up the use of clean energy and providing energy for growth. An underlying theory of change within DfID's research investments links the proposed research activities and outputs to expected outcomes and their impacts. There remain, however, some questions over how those impacts (differently defined for different parts of DfID's research programme) might best be measured and progress evaluated. Some measurements draw on common generic research deliverables, each with reportable indicators, including: academic research publications aiming to improve understanding of energy options and opportunities; specific publications aiming to improve operational understanding and local context; proposals for improving local capacity and skills, including in the area of research; and proposals for improving access to knowledge.

Research programmes more focused on innovation may also include deliverables targeting technologies supported through the different stages of technology readiness, and business models that are tested and scaled up. Anticipated outcomes that capture the potential contribution to evidence-based policymaking include the cumulative number of enabling policy and environment examples informed by the research in target countries – energy regulations, legal measures, adopted reforms, policy guidelines, frameworks, tools, models – and relevant projects and proposals reflecting and incorporating research findings. Other DfID energy research programme developments include the commissioning of a number of increasingly large-scale programmes of energy research, which illustrate evolving thinking on the management and delivery of research, the contribution potential of different stakeholders and the processes of delivering research impact (for examples of such projects see https://www.lcedn.com/initiatives/category/USES-Network)

DfID's rapidly growing applied research programme in this area has been characterised by changes in terms of the researchers involved, greater local engagement and more demanding reporting requirements, reflecting the rapidly evolving nature of the sector and funders' expectations. The programme is characterised by a broadening of the dimensions of research partnerships, a tighter definition of research scope, and integration of innovation and business support, different forms of commissioning of research, co-funding and a growing attention to research impact.

A key dimension to this evolution has been an increasing engagement with the academic community, in the UK and internationally, as well as with a range of Southern partners, civil society and philanthropic foundations, innovators and businesses, and investors. Concerns about climate and about growing green markets have led to exploring the best ways of supporting the take-up and adoption of promising technologies and business models, and creating conducive eco-systems in the UK and developing countries. The shift to a more applied focus to DfID-supported research and innovation work has brought in a wider range of research partners, including innovators and business incubators, facilitated by applying the theory of change approaches (Vogel, 2012). More attention has been paid to the importance of local energy contexts and social science perspectives, although a considerable need remains for further engagement with, and prioritisation of, the social sciences. The importance of broader and more longitudinal research activity remains. Collaborative partnerships with the relevant research councils, including through the Global Challenges Research Fund, could provide an important basis for combining energy-related applied research and more academic research as well as a basis for effective networking to deliver coherence and value for money, although the prospects for such collaboration at least in the energy sector currently remain unclear, and the spaces for qualitative social science analysis of contexts affecting low carbon transitions in the communities of the poor and marginal are few and far between as compared to technically driven research projects.

The network approach to measuring impact: the USES programme

Alongside the UK government's growing international involvement in tackling international energy poverty, recent years have seen a surge of interest in research on the imbricated themes of climate change, energy access issues and low carbon transitions among academic organisations. The LCEDN was launched by UK researchers in 2012 to respond to this interest and the demand for interaction and cooperation on research for low carbon development described above, bringing together the diverse and multi-disciplinary UK academic community working on energy and development issues and building relationships with business, public sector and civil society organisations in the same fields. Initially centred on five academic research centres, it has expanded to include practitioners, civil society, the private sector and the policy community. The network aims to pinpoint UK strengths, identify where they could best be deployed and highlight areas where expertise needs further development, specifically by identifying commonalities of interest and encouraging disparate stakeholders with little or no knowledge of each other's work and interested to learn each other's language as a first step to cooperation.

The Understanding Sustainable Energy Solutions (USES) research programme (jointly funded by EPSRC, DFID and DECC), which was initiated in 2013 has grown into a subsidiary network assemblage of the LCEDN, dedicated to providing a platform for deepening the work of the LCEDN and its interactions with DfID, increasingly through a growing body of international partners. The key research themes of the USES programme (https://

www.lcedn.com/initiatives/category/USES-Network) emerged out of conversations between funding bodies, brokered via their interactions at LCEDN management committee meetings, and discussions initiated with the wider research community at LCEDN events. The overall intention of the programme was:

> to increase clean energy access, resilience and wealth creation in developing countries (particularly for the urban and rural poor), through high quality research that improves the understanding and evidence-base of opportunities and challenges associated with clean energy for development. (EPSRC/DfID/DECC, 2012, p. 1)

The development of the programme was explicitly connected to the UK government's recognition of the growing urgency of tackling climate change and the need to contribute to a longer-term goal of increased clean energy access, resilience and wealth creation for low-income households in developing countries. The overall objectives of the programme, focusing on five themes – energy systems and decentralised use, solar, bioenergy, urban and transport, and energy efficiency – were that it would lead to: improved understanding of clean energy options and opportunities for developing countries; improved understanding of the social, market and political economy aspects of scaling sustainable energy access for poor people; strengthened developing country research capacity on clean energy and improved access to practical and policy-relevant knowledge on the challenges and opportunities for sustainable energy solutions in developing countries. The programme funded 13 projects, each of which involved research taking place in more than one country. Each project brought together researchers from a variety of different disciplinary backgrounds and involved collaborations beyond the academic community through partnerships with NGOs, policymakers and/or private sectors actors.

The programme provided an opportunity to develop a practical platform for encouraging dialogue between the UK academic community and DfID's research and evaluation division. Alongside the individual projects, the LCEDN provided a networking platform for the researchers to discuss experiences and explore synergies and cross-cutting themes. The LCEDN's original role in networking the projects for EPSRC was expanded via DfID funding to take on a greater technical monitoring and reporting role, promote the means and measurement of research outcomes and impact and provide opportunities for joint activities designed to engage with stakeholders and enhance the take-up of the lessons learnt across the programme as a whole.

The USES programme highlighted different expectations and approaches to research impact. Measurements of impact needed to capture the interest of the research councils in academic output and impact case studies, as well as the interest of DfID in developmental outcomes. However, initial reporting requirements tended to focus on relatively short-term impacts and did not reflect the wider understanding of networked impact outlined in this paper. These differential definitions and expectations of impact were themes that resurfaced fairly frequently during ongoing discussions of impact among the USES network partners, which provided a number of useful lessons on the dimensions and contours of research impact.

The relatively short-term nature of the projects, a common critique by researchers and practitioners, and the lack of channels for follow-up projects meant that potential impact through capacity building and network establishment frequently could not be fully realised. One USES project observed a major shift in perspective among policy actors, but identified as a major risk that 'once funding ceases, stakeholders go back to their routine'. Others pointed out that the need to devote funding to engagement and policy uptake activities had too often been underplayed in project formulation and implementation and had required additional expenditure. Some trade-offs were also identified between different measures of impact required by different funders. For example, one USES project reported that:

> some of the journal papers have been published in less prestigious journals than the administrative PI would normally aim for but it is hoped that these will be of more relevance to Southern partners, and they have to date generated significant implementation interest from a variety of stakeholders. (personal communication)

The USES projects produced a diverse range of outputs including academic journal articles (both top-rated internationally leading journals and the more practically focused, frequently regional, journals mentioned in the comments above), books, book chapters, policy briefs, apps, economic models, data sets, toolkits, decision support tools and videos. Greater training and support for researchers would enhance the production of non-academic outputs and other forms of media to engage other sectors and to disseminate the outcomes of the research more widely. USES network meetings provided opportunities for cross-fertilisation of successful impact strategies but the need was identified for better capture and communication of collective achievements and results against anticipated outcomes and impact indicators.

The experiences of the USES projects showed that the dialogical production of impact depends heavily upon time, resources and wide stakeholder involvement. It takes time to build relationships and platforms, to share knowledge and develop impact, and it is these relational conditions of possibility for an impact that need much better recognition in participatory ownership of project objectives. Research does not automatically lead to impact, so considerable resources need to be devoted to creating opportunities, nurturing relationships and influencing policy strategically. The diversity of voices requires the identification of stakeholders and facilitation of cross-sectoral understandings. For example, one USES project reported:

> The ... field is complex and highly political. Relationship building is therefore important but also a slow process. One realisation ... is the need to be opportunistic in developing win-win partnerships.

Using these learnings, and drawing on the support of the LCEDN, DfID has developed a series of platforms where the uptake of learning from the UK research programme can be better coordinated and lessons shared. This has now been taken forward more systematically in the new Transforming Energy Access (TEA) programme. The TEA programme provides an example of how different partners and approaches are being integrated into a multi-layered programme aiming to increase the use of affordable decentralised clean energy options for poor households and enterprises, through a combination of innovative technologies, delivery models and financing, and complementary capacity building, providing for a particular focus on research uptake and impact measurement. Like USES, TEA is intended to embrace a wide range of stakeholders including researchers from a variety of disciplines, innovators and entrepreneurs, social impact investors, utility and public sector employees, and evaluators often with differing perspectives and ways of working. The overall TEA programme objective is

> [to] deliver new technologies and robust evidence on the critical barriers hampering systemic change and scaling up energy access, working with Southern researchers and entrepreneurs to drive locally relevant innovation and delivery. (DfID, 2016b, p. 2)

Taken as a whole, TEA is expected to have a transformative impact on the design and deployment of renewable energy solutions across the Global South, especially in Africa. The LCEDN's programme of support builds on the experiences of the USES programme and focuses on capacity building around a number of key themes essential for achieving the goals of the TEA, namely: gender, governance, value chains, innovative forms of finance and delivery models, waste and sustainability, approaches towards innovation, impact assessment methodologies, and transdisciplinary and cross-sectoral working. The intention is to help integrate and strengthen the wider UK energy and development research and innovation delivery context, including working with other major initiatives such as DfID's Energy and Economic Growth programme, the EU Africa Energy Partnership and other initiatives.

Conclusion

This paper has explored briefly the different ways in which impact is conceived by the diverse set of actors involved in research and innovation around low carbon energy and international development in the UK, and the ways in which conversations on impact are being undertaken to improve the value of the term to funders, as well as measuring real change in beneficiary communities. The experience of DfID's USES Programme has shown that measuring and improving impact in relation to energy trilemma initiatives requires careful re-consideration and review of these diverse needs. The programme also demonstrated the need to recognise and devote specific resources to achieving longer-term actor–network dialogues and capacity-building impacts beyond individual projects, and using these to build a more networked approach to strengthening impact. These are issues that need to be taken into account when judging the impact of interventions funded under the RCUK's GCRF programme, and that require social science intelligence to temper lingering technology transfer models of how development happens.

It is interesting to observe that TEA also contains a component designed to explore opportunities to take forward follow-up learnings from the USES programme and others designed to nurture relationships and knowledge sharing at a regional level. For example, the TEA programme start-up phase includes a set of coordinated actions in East Africa, with actions focused on trailing innovative ways to disseminate academic outputs with targeted support, contributing to workshops exploring the role of governance with policymakers in the region as well as a final workshop focused on the USES projects working with local communities. The USES network hosted several discussions around different conceptualisations of impact that have fed into the design of a workstream under the TEA initiative revolving around detailed assessments of the existing relationships between researchers, policymakers and other stakeholders within specific national contexts (initially Kenya and Bangladesh), the assessment of knowledge needs and identification of communication channels and development of joint impact planning activities across different sectors.

The evolution of a whole system approach to low carbon energy provision and integrated energy research programmes such as the TEA highlight the potential to build such networks. As the UK energy research programmes develop, and increased attention is paid to strategic partnerships and research impact, the kind of networking and outreach promoted by the LCEDN over the past five years can be seen increasingly as an important means of bringing different communities and interests together, and building relationships between UK policymakers, academia, entrepreneurs and business (while acknowledging their mutual frictions). It is only through the establishment of these long-term relationships and coordinated programmes that the wider less linear and more emergent versions of impact and innovation for SDGs can be nurtured and developed in response to the articulated perceptions of energy needs of poor and marginalised women and men (Ely, Smith, Leach, Stirling, & Scoones, 2013). These insights have been incorporated into the initial fast-track activities for the TEA programme, but further work is needed on refining research impact measurements, which go beyond reporting and help build collaboration.

Acknowledgements

The views expressed in this paper are solely the responsibility of the authors and do not necessarily reflect the views of the institutions funding the work.

Disclosure statement

No potential conflict of interest was reported by the authors.

Funding

This work was supported by the Engineering and Physical Sciences Research Council (EPSRC) [grant number EP/L014858/1] and Department for International Development (DfID) [grant number GB-1-203781] (for details of the USES programme, see the website maintained by the LCEDN: http://www.lcedn.com/uses). The views expressed in this paper are solely the responsibility of the authors and do not necessarily reflect the views of the institutions funding the work.

Notes on contributors

Ed Brown is Senior Lecturer in Geography at Loughborough University, UK. He is an economic/political geographer who has worked on a range of development issues over the past 25 years including questions of governance, transparency, corruption and equity. His recent research focuses on the interplay of energy and international development issues. He is the Co-Chair of the UKLCEDN. He has received research funding from EPSRC, DfID and Innovate, among others, for his work in this area.

Ben Campbell is Senior Lecturer in the Department of Anthropology at Durham University, UK, and director of its MSc in Sustainability, Culture and Development. He is Co-Chair of the LCEDN. His research focuses on poverty, livelihoods and resources in Nepal, on innovation among mountain indigenous communities and on the impacts of environmental conservation measures on local institutions of common property management. With the support of the Durham Energy Institute, he has studied energy transitions in Nepal from fuelwood use to biogas for SMEs at high altitudes. His publications include *Living between juniper and palm: Nature, culture and power in the Himalayas* (Oxford University Press, 2013).

Jon Cloke is National Network Manager for the LCEDN and is based in the Geography Department at Loughborough University. He works at mapping the UK low carbon for development research landscape and coordinating the projects funded under the DfID/EPSRC/DECC-funded Understanding Sustainable Energy Solutions (USES) research programme. He is also a researcher on the LCEDN's programme of support for DfID's Transforming Energy Access (TEA) programme and the Solar Nano-grids (SONG) USES project. He has worked as a Consultant to various NGOs, including the UK-based NGO Practical Action, developing policy documents for guiding strategy.

Long Seng To is a Research Associate with the LCEDN in the Department of Geography at Loughborough University. Her research focuses on sustainable energy and development in Asia and Africa. She is particularly interested in the role of capacity development and agro-industries on expanding energy access, and in ways of enhancing community resilience using renewable energy. She has also worked as a consultant to the Australian government on climate change and energy policy, and with non-government organisations to support youth-led sustainable development initiatives. She is a trustee of Engineers Without Borders UK.

Britta Turner is a Research Associate with the LCEDN and is based in the Department of Anthropology at Durham University. Her doctoral research focused on the issues surrounding the use of photovoltaic technology in domestic settings in the UK and Sri Lanka. Her PhD explored the framing of domestic solar as a particular socio-material assemblage within both country contexts and illustrated how solar electricity acts not as a neutral resource, which is handled and interpreted by human beings, but rather as a spatially and temporally diverse force with properties and propensities that encourage particular orderings of meaning and matter.

Alistair Wray worked for the UK's DfID until 2017. He is a civil engineer by training and has worked with consulting firms and at a university development centre prior to joining DfID in 1991. At DfID he held a variety of posts working on both policies relating to infrastructure, and country and regional programmes, including East Asia and Southern Africa. He was Senior Energy Advisor in DfID's Research and Evidence Division with responsibility for the USES programme and the development of DfID's new Transforming Energy Access initiative.

References

Alvarez, S., Douthwaite, B., Thiele, G., Mackay, R., Córdoba, D., & Tehelen, K. (2010). Participatory impact pathways analysis: A practical method for project planning and evaluation. *Development in Practice*, *20*(8), 946–958. doi:10.1080/09614524.2010.513723.

Bazilian, M., & Pielke, R. (2013). Making energy access meaningful. *Issues in Science and Technology*, *29* (4), 74–78. Retrieved from http://issues.org/29-4/making-energy-access-meaningful/

Bray, F. (2007). Gender and technology. *Annual Review of Anthropology*, *36*, 37–53. doi:10.1146/annurev.anthro.36.081406.094328

British Academy. (2016). *Crossing paths: Interdisciplinary institutions, careers, education and applications*. Retrieved from http://www.britac.ac.uk/sites/default/files/Crossing20Paths20-20Full20Report_2.pdf

Brown, E., Cloke, J., Gent, D., Johnson, P., & Hill, C. (2014). Green growth or ecological commodification: Debating the green economy in the global south. *Geografiska Annaler: Series B, Human Geography*, *96*, 245–259. doi:10.1111/geob.12049

Bulkeley, H., & Castan Broto, V. (2013). Government by experiment? Global cities and the governing of climate change. *Transactions of the Institute of British Geographers*, *38*(3), 361–375.

Byrne, R., Schoots, K., Watson, J., Ockwell, D., de Coninck, H., Gallagher, K. S., & Sagar, A. (2012). Innovation systems in developing countries. *CDKN Climate Development Technology Policy Brief*. SPRU, University of Sussex. Retrieved from http://unfccc.int/ttclear/misc_/StaticFiles/gnwoerk_static/TEM_tec_cfi_ee/7843d4ba5e5e459c99deb4e47b972e83/96d0add518b44ebbae1b977e1ef3dc65.pdf

Castree, N., Adams, W., Barry, J., Brockington, D., Buescher, B., ... Corbera, E. (2014). Changing the intellectual climate. *Nature Climate Change*, *4*, 763–768.

Cook, P. (2011). Infrastructure, rural electrification and development. *Energy for Sustainable Development*, *15*, 304–313. doi:10.1016/j.esd.2011.07.008

de Laet, M., & Mol, A. (2000). The Zimbabwe bush pump: Mechanics of a fluid technology. *Social Studies of Science*, *30*(2), 225–263.

DfID. (2014). *Summary plan and priorities: Clean energy and innovation*. Internal DfID discussion document. DfID Research and Evidence Division.

DfID. (2016a, October). *DfID research review*. Retrieved from https://www.gov.uk/government/publications/dfid-research-review

DfID. (2016b, June). *Applied research programme on transforming energy access (TEA): Terms of reference for a research programme delivery consortium*. DfID Research and Evidence.

DfID. (2017, January 31). *Economic development strategy: Prosperity, poverty and meeting global challenges*. DfID. Retrieved from https://www.gov.uk/government/uploads/system/uploads/attachment_data/file/587374/DfID-Economic-Development-Strategy-2017.pdf

Dove, M., & Kammen, D. (2015). *Science, society and the environment: Applying anthropology and physics to sustainability*. London: Routledge.

Ely, A., Smith, A., Leach, M., Stirling, A., & Scoones, I. (2013). Innovation politics post-Rio+20: Hybrid pathways to sustainability? *Environment and Planning C: Government and Policy*, *31*, 1063–1081. doi:10.1068/c12285j.

EPSRC/DfID/DECC. (2012). *Energy and international development: Understanding sustainable energy solutions. Call for expressions of interest*. Retrieved from https://www.epsrc.ac.uk/files/funding/calls/2012/energy-and-international-development/

Feldman, S., & Briggs, S. (2012). The politics of international assessments: The IAASTD process, reception and significance. *Journal of Agrarian Change*, *12*(1), 144–169. doi:10.1111/j.1471-0366.2011.00333.x

Leach, M., Scoones, I., & Wynne, B. (Eds.). (2005). *Science and citizens: Globalization and the challenge of engagement*. London: Zed Books.

Makhijani, S. (2014). *Fossil fuel exploration subsidies: United Kingdom*. Country Study. ODI. Retrieved from www.odi.org/sites/odi.org.uk/files/odi-assets/publications-opinion-files/9245.pdf

Mission Innovation. (2017). *United Kingdom (Description of the UK's programme involvement in the Mission Innovation programme)*. Retrieved from http://mission-innovation.net/participating-countries/united-kingdom/

Nicholls, A. (2015). *Synthetic grid: A critical framework to inform the development of social innovation metrics* (CRESSI Working Papers, No. 14/2015). ISBN: 978-0-9933861-3–8.

Ockwell, D., & Byrne, R. (2016). *Sustainable energy for all: Innovation, technology and pro-poor green transformations*. Abingdon: Routledge.

Reale, E., Avramov, D., Canhial, K., Donovan, C., Flecha, R., ... Holm, P. (2017). A review of literature on evaluating the scientific, social and political impact of social sciences and humanities research. *Research Evaluation*, 1–11. doi:10.1093/reseval/rvx025.

Richards, P. (1985). *Indigenous agricultural revolution: Ecology and food production in West Africa*. London: Unwin Hyman.

Robbins, P., Wield, D., & Wilson, G. (2017). Mapping engineering and development research excellence in the UK: An analysis of REF2014 impact case studies. *Journal of International Development*, *29*(1), 89–105. doi:10.1002/jid.3255

Rolffs, P., Ockwell, D., & Byrne, R. (2015). Beyond technology and finance: Pay-as-you-go sustainable energy access and theories of social change. *Environment and Planning A*, *47*(12), 2609–2627. doi:10.1177/0308518X15615368

Shortt, N., Pearce, J., Mitchell, R., & Smith, K. (2016). Taking health geography out of the academy: Measuring academic impact. *Social Science and Medicine*, *168*, 265–272. doi:10.1016/j.socscimed.2016.06.048

Smith, K. E., & Stewart, E. (2017). We need to talk about impact: Why social policy academics need to engage with the UK's research impact agenda. *Journal of Social Policy*, *46*(1), 109–127. doi:10.1017/S0047279416000283

Strathern, M. (Ed.). (2000). *Audit cultures: Anthropological studies in accountability, ethics and the academy*. London: Routledge.

Tomei, J., & Gent, D. (Eds.). (2015). *Equity and the energy trilemma: Delivering sustainable energy access in low-income communities*. London: International Institute for Environment and Development.

UKCDS. (2014). *Mapping infrastructure research for development*. UKCDS. Retrieved from http://www.ukcds.org.uk/sites/default/files/content/resources/UKCDS_Mapping_Infrastructure_Research_for_Dev_May_2014_0.pdf

Vogel, I. (2012). *Review of the use of 'Theory of Change' in international development*. UK Department of International Development. Retrieved from http://www.theoryofchange.org/pdf/DFID_ToC_Review_VogelV7.pdf

Wellberry, D. (2009). The general enters the library: A note on disciplines and complexity. *Critical Enquiry*, *35*(4), 982–994. doi:10.1086/599588

Winter, T., Matinga, M. N., Ulsrud, K., & Standal, K. (2017). Women's empowerment through electricity access: Scoping study and proposal for a framework of analysis. *Journal of Development Effectiveness*, *9*, 389–417.

Winther, T. (2015). Impact evaluation of rural electrification programmes: What parts of the story may be missed? *Journal of Development Effectiveness*, *7*(2), 160–174. doi:10.1080/19439342.2015.1008274

World Energy Council. (2013). *Balancing the 'energy trilemma'*. Retrieved from https://www.worldenergy.org/wp-content/uploads/2013/09/Trilemma-original.pdf

Index

Note: Page numbers in *italics* refer to figures
Page numbers in **bold** refer to tables

For Product Safety Concerns and Information please contact our EU representative GPSR@taylorandfrancis.com Taylor & Francis Verlag GmbH, Kaufingerstraße 24, 80331 München, Germany

Batch number: 10726294

Printed by Printforce, the Netherlands